The Logic of

The Logic of Decision

SECOND EDITION

Richard C.
Jeffrey

University of Chicago Press
CHICAGO AND LONDON

Richard C. Jeffrey is professor of philosophy at
Princeton University.

The University of Chicago Press, Chicago 60637
The University of Chicago Press, Ltd., London

Printed in the United States of America
90 89 88 87 86 85 84 83   1 2 3 4 5

Library of Congress Cataloging in Publication Data

Jeffrey, Richard C.
The logic of decision.

Includes bibliographical references and index.
1. Bayesian statistical decision theory. I. Title.
QA279.5.J43 1983   519.5′42   82-13465
ISBN 0-226-39581-2

Uxori delectissimae Laviniae

Contents

Preface

This book uses elementary logical and mathematical means to a philosophical end: elucidation of the notations of *subjective probability* and *subjective desirability* or *utility* which form the core of Bayesian decision theory.

Most of the book (chapters 4–10) is devoted to a new theory of preference between (truth of) propositions, within which a Bayesian agent's probability and utility functions are determined by features of his preference ranking. Here the elementary logical operations on propositions (denial, conjunction, disjuction) do the work which is done by the operation of forming gambles in the "classical" theory of Ramsey and Savage. Classically, the agent's preference ranking of gambles determines his utility function up to a linear transformation with positive coefficient, and determines his probability function completely; but here (see chapter 6), the preference ranking of propositions determines the utility function only up to a fractional linear transformation with positive determinant, and determines the probability function only to within a certain quantization. The classical case is obtained here if the preference ranking is of the sort that can only be represented by a utility function that is unbounded both above and below; and it is shown (chapter 10) that the present theory is immune to the St. Petersburg paradox, so that one can reasonably be a Bayesian in the present sense and still have an unbounded utility function.

At each point I have tried to keep the mathematics as simple as possible: interest, persistence, and a high school education are the only prerequisites for reading the book, except for bits toward the end. Chapters 4 and 5 will provide the reader with the necessary logic and probability theory, and the first two chapters do the same for decision theory. Exercises of various degrees of difficulty follow each of the first eight chapters. Some of the exercises relate directly to points made in the text; others introduce new points.

The last two chapters treat philosophical questions that have nearly the same form in the new theory as in the classical theory and which are peculiarly relevant to the subjectivistic account of probability. Chapter 11 deals with uncertain evidence, where an observation leads the agent to change his degrees of belief in one or more propositions to new values that fall short of 1. It is argued that the "evidence" of the senses can be accounted for in these terms without supposing that the agent is aware of sense data as such. Chapter 12 suggests how notions of objective probability and desirability might fit into the subjectivistic framework and has something to say about inductive reasoning.

In this second edition, chapter 9 is quite new: an exposition of Ethan Bolker's axiomatization of my system replaces a very different chapter in the first edition, which the preface said could "be profitably omitted on any but a very close reading." Section 1.7 is also new: a modification of the rule "Choose an act of maximum estimated desirability" that agrees with that rule in ordinary cases, but that gives what I take to be the right solutions in many (alas! not all) of the bothersome cases where agents see their decisions as merely symptomatic of states of affairs that they would promote or prevent if they could. "Preference among Preferences" is reprinted here as the Appendix, much as it appeared in the *Journal of Philosophy* in 1974. In response to suggestions of Sandy Zabell's, sections 7.5 and 7.6 have been redone in order to clarify the proof of (7-5). The rest is as in the first edition, except for minor corrections, some new problems (in section 1.6), and reviews of developments over the past sixteen years or so in the new Notes and References at the ends of chapters.

The work reported in the first edition was mostly done at Stanford University and the Institute for Advanced Study in 1961–1964. But chapter 11 and section 12.4 derive from my Ph.D. dissertation (Princeton, 1957), which was done under the supervision of C. G. Hempel and, sometimes, Hilary Putnam. It was in Hempel's seminar that I found out about Ramsey and Savage and got interested in decision theory. (At Chicago, 1946–51, I had learned about confirmation theory from Carnap.) I went to Stanford because of the interest there in probability and decision theory, and the inception of this book owes much to the stimulation of Donald Davidson and Patrick Suppes there (see sections 3.9 and 10.7 for some details).

Important parts of this book are mainly accounts of mathematical discoveries of Ethan Bolker's. I was able to finish my project only when Bolker and I saw that his mathematical theorems applied to my theory of preference among propositions. We met and made that discovery just as Kurt Gödel independently noticed a part of it: he saw that the right transformations in the equivalence theorem must be those specified here in section 6.1, and he conjectured that some such uniqueness theorem as that in chapter 8 could be proved by linear algebra. But Bolker had already proved the equivalence and uniqueness theorems and had proved an existence theorem which he easily applied to my theory of preference to get the axiomatization reported in chapter 9. If the system is applied mathematics, the mathematics is Bolker's and the application mine. (Chapters 5 and 7 report mathematics that I had worked out on my own, before meeting Bolker—but I had supposed that by working harder I could discover how preferences determine utilities up to a positive linear transformation and determine probabilities uniquely. Bolker's uniqueness theorem showed that these determinations are possible if, and only if, the preference ranking determines a utility function that is unbounded both above and below.)

At the early stages of my investigations, I got much help and useful criticism from Rudolf Carnap, Donald Davidson, L. J. Savage, and Patrick Suppes, and more later (while preparing the first edition) from Paul Benacerraf, Gilbert Harman, Robert Nozick, and Howard Smokler. In the treatment of Ramsey's system, I profited from reading John M. Vickers' Ph.D. thesis (Stanford, 1962), as well as from the borrowings from Davidson and Suppes noted in sections 3.9 and 10.7. Thanks are also due to Vickers for his careful reading of the manuscript of the first edition and for his valuable criticism of it. Many students at Stanford and Princeton Universities and at the City College of New York have helped by locating errors and obscurities in earlier versions of this material. Ethan Bolker, Henry Kyburg, and Frederick Schick have pointed out obscurities and flaws in the first edition that I think have been corrected here. Schick pointed out a reparable flaw in the old chapter 9 (see the *Journal of Philosophy* 64 [1967]: 396–401). Both Bolker and Kyburg faulted the uniqueness proof in chapter 8 of the first edition as insufficiently explicit, given what they took to be its prima facie unworkability in the absence of an assumption of countable additivity. I have now made the proof plainer, I think, and have become persuaded of its correctness with the help of coaching from Persi Diaconis, David Freedman, and Sandy Zabell. To Sandy Zabell I am indebted for much more: he has read the whole manuscript carefully, identified many errors, and made detailed suggestions for their correction and for the inclusion of other material, for example, explanations and references.

Thanks are due to the National Science Foundation, the John Simon Guggenheim Memorial Foundation, and the National Endowment for the Humanities for support of my work on decision theory and allied matters since publication of the first edition of this book, and to the Air Force Office of Scientific Research for support of earlier parts of the work at

Stanford and the Institute for Advanced Study. (In the aftermath of Sputnik, and well into the sixties, the Air Force Office of Scientific Research supported pure, "blue skies" research in a disinterested way.) Thanks are also due to Sandra Peterson for getting me to quit smoking the first time, to the American Cancer Society and the Consumers Union for permission to reproduce the data on smoking and death in problem 4 of section 1.6, and to the editors of the *Journal of Philosophy* for permission to reprint "Preference among Preferences."

For sentimental reasons among others, it is a pleasure to have this book published by the University of Chicago Press, which published many of the works of my teacher Rudolf Carnap. It was in Carnap's seminars that I met my friend and tutor, Abner Shimony, whose sympathetic comments on the work in progress were most welcome. The first edition was dedicated to Carnap, and the last few works of the first edition (of section 12.8 here) were included out of love and respect for him and his vision of things, i.e.,the *Aufklärung*, in the colors in which it dawned upon me when I was seventeen. In the same sprit—mutatis mutandis—this edition is dedicated to my wife, Edith.

1

Deliberation
A Bayesian Framework

To judge what one must do to obtain a good or avoid an evil, it is necessary to consider not only the good and the evil in itself, but also the probability that it happens or does not happen; and to view geometrically the proportion that all these things have together.

The Port-Royal Logic, 1662

To deliberate is to evaluate available lines of action in terms of their consequences, which may depend on circumstances the agent can neither predict nor control. The name of the eighteenth-century English clergyman Thomas Bayes has come to be associated with a particular framework for deliberation, in which the agent's notions of the probabilities of the relevant circumstances and the desirabilities of the possible consequences are represented by numbers that collectively determine an estimate of desirability for each of the acts under consideration. The Bayesian principle, then, is to *choose an act of maximum estimated desirability*. (*An* act rather than *the* act, since two or more of the possible acts may have the same, maximum estimated desirability.) The numerical probabilities and desirabilities are meant to be subjective in the sense that they reflect the agent's actual beliefs and preferences, irrespective of factual or moral justification.

1.1 Acts, Conditions, Consequences

In the simplest cases the number of possible acts that the agent believes are available to him is finite, as is the number of possible circumstances that he regards as relevant to the outcomes of the acts. The first stage of deliberation is then an analysis of the situation into *acts, conditions,* and *consequences* which can be summarized by a *consequence matrix* as in the following two examples.

Example 1: Swimming

A ticket allowing the bearer to use a certain beach all weekend costs three dollars if purchased during the week, while a single day's admission costs two dollars if paid on the day. The consequence matrix might be as follows.

	0 days of good weather	1 day of good weather	2 days of good weather
Buy a weekend ticket	Pay $3 for 0 days of swimming	Pay $3 for 1 day of swimming	Pay $3 for 2 days of swimming
Pay admission daily	Pay $0 for 0 days of swimming	Pay $2 for 1 day of swimming	Pay $4 for 2 days of swimming

Example 2: Nuclear disarmament

A crude version of one sort of argument for nuclear disarmament might use the following consequence matrix.

	War	Peace
Arm with nuclear weapons	Extinction of human life	Continuation of life under present conditions
Disarm	Continuation of life under abhorrent conditions	Golden age

In each case, the row headings of the consequence matrix represent possible acts, and the column headings represent possible conditions which might affect the outcomes of the acts. The consequences described by the entries in the matrix need not be new entities, different in kind from the acts and conditions which produce them. Thus in the swimming example, the act of buying a weekend ticket is performed by paying three dollars to the right person at the right time. Therefore, the descriptions of the consequences in the upper row of the matrix ("Pay $3 for . . .") are, in part, descriptions of the act.

1.2 Desirabilities and Probabilities

Entries in consequence matrices are best considered as notes made by the deliberating agent to help determine the numerical desirabilities of situations which he expects to arise if he performs one or another act under various conditions. The desirabilities are, if you like, desirabilities of *consequences,* but a description of the consequences of a certain act under a certain condition need be nothing more than a joint description of the act and the condition, as the following example suggests.

Example 3: The right wine

The dinner guest who is to provide the wine has forgotten whether chicken or beef is to be served. He has no telephone, has a bottle of white and a bottle of red, and can only bring one of them (in an oversized pocket) since he is going by bicycle. The consequence matrix might well be the following.

	Chicken	Beef
White	White wine with chicken	White wine with beef
Red	Red wine with chicken	Red wine with beef

Of course he might then modify the entries as a step toward formulating numerical desirabilities, perhaps (idiosyncratically) as follows.

	Chicken	Beef
White	The right wine	The wrong wine
Red	An odd wine	The right wine

On the other hand, he might have gone directly from the original consequence matrix to (say) the following numerical *desirability matrix*.

	Chicken	Beef
White	1	−1
Red	0	1

Suppose that the guest regards the two possible conditions as equally likely, regardless of whether he brings white wine or red. Then the probabilities are as in the following *probability matrix*.

	Chicken	Beef
White	.5	.5
Red	.5	.5

Now given the numerical probabilities and desirabilities, we can estimate the desirability of each act by multiplying corresponding entries in the probability and desirability matrices and then adding across each row.

Dropping the row and column headings, the matrices are

.5	.5
.5	.5

1	-1
0	1

Multiplying corresponding entries, we obtain a new matrix,

(.5)(1)	(.5)(-1)
(.5)(0)	(.5)(1)

=

.5	$-.5$
0	.5

Finally, adding across each row, we have

$$(.5) + (-.5) = 0$$

as the desirability of the first act (white), and

$$0 + .5 = .5$$

as the desirability of the second act (red). Then, bringing red wine has the higher estimated desirability, and according to Bayesian principles, is the better choice.

The two rows of the probability matrix need not be identical; in other words, the probabilities of the conditions need not be independent of the acts.

Example 4: Dependence

The guest might think it possible that the meat will be chosen to suit the wine; perhaps the cooking is to be done after he arrives. If he is convinced that this will be done, his probability matrix will be

1	0
0	1

Multiplying by corresponding entries in the desirability matrix and adding across rows would yield

$$(1)(1) + (0)(-1) = 1$$

and

$$(0)(0) + (1)(1) = 1$$

as estimated desirabilities of the two acts. Then according to Bayesian principles, he should consider it unimportant which bottle he brings.

Example 5: Weaker dependence

Suppose the agent thinks it likely as not that the meat will be chosen to suit the wine. Then he might combine the probability matrices

1	0
0	1

.5	.5
.5	.5

by averaging corresponding entries to obtain

.75	.25
.25	.75

as his over-all probability matrix. This would yield estimated desirabilities

$$(.75)(1) + (.25)(-1) = .5$$

$$(.25)(0) + (.75)(1) = .75$$

for the two acts, so that again, the second is better.

1.3 Summary and Rationale

A formal Bayesian decision problem is specified by two rectangular arrays (matrices) of numbers which represent probability and desirability assignments to the act-condition pairs. The columns represent a set of incompatible conditions, an unknown one of which actually obtains. Each row of the desirability matrix,

$$d_1 \qquad d_2 \qquad \ldots \qquad d_n$$

represents the desirabilities that the agent attributes to the n conditions described by the column headings, on the assumption that he is about to perform the act described by the row heading; and the corresponding row of the probability matrix,

$$p_1 \qquad p_2 \qquad \ldots \qquad p_n ,$$

represents the probabilities that the agent attributes to the same n conditions, still on the assumption that he is about to perform the act described by the row heading. To estimate the desirability of the act, multiply corresponding probabilities and desirabilities, and add:

$$p_1 d_1 + p_2 d_2 + \ldots + p_n d_n .$$

Having done this for each row, select for performance one of the acts for which this sum of products is greatest.

The rationale behind the procedure will be familiar to any gambler, and to any actuary: if there are n possible events, of which just one will take place, and if the probabilities of these events are

$$p_1 \quad p_2 \quad \ldots \quad p_n$$

and if you stand to gain amounts

$$d_1 \quad d_2 \quad \ldots \quad d_n$$

of dollars if the corresponding events actually occur, then the actuarial value of the gamble is the sum of products

$$p_1 d_1 + p_2 d_2 + \ldots + p_n d_n \, .$$

This actuarial value is the fair price of a lottery ticket on which the payoff will be d_1 or d_2 or . . . or d_n with probabilities $p_1, p_2, \ldots, p_n$.

Example 6: Coin tossing

There are three possible events, 0, 1, and 2 heads, as outcomes of two good tosses of a normal coin. The probabilities of the respective outcomes are .25, .5, .25. If in the three events the player stands to gain 50¢, 20¢, and nothing, the actuarial value of the gamble is

$$(.25)(50) + (.5)(25) + (.25)(0) = 25¢ \, ,$$

which is thus the fair price of a ticket allowing the bearer to play.

Example 7: It takes 50¢ to ride the subway

With only 35¢ in his pocket, the agent urgently needs to use the subway, which costs 50¢. Under the circumstances, he might be well advised to pay 35¢ (more than the fair price) for the gamble in example 6, in order to raise his probability of getting onto the subway from 0 to .25. In particular, suppose that the desirability of 50¢ is some positive number d, but that the desirability of any smaller amount is 0. Then the desirability of the gamble is

$$.25d + (.5)(0) + (.25)(0) = .25d \, ,$$

which, being positive, is greater than the desirability (0) of the 35¢ the agent has in his pocket. Then the act of exchanging the 35¢ for the gamble is preferable to the act of keeping the 35¢.

1.4 Incompletely Specified Desirabilities

The (estimated) desirability of an act is its actuarial value, measured in units that may, but need not, correspond to dollars. The swimming example is fairly representative of a class of decision problems in which monetary considerations can plausibly be used to obtain numerical desirabilities. The agent might reason that the prospects of 0, 1 and 2 days' swimming have unknown monetary values of x, y, and z dollars, and he might proceed to identify the desirability of each consequence with the value of the amount of swimming in question minus the amount spent for tickets. He would then obtain the following desirability matrix:

$x-3$	$y-3$	$z-3$
x	$y-2$	$z-4$

Example 8: Solving the swimming problem

To put the matter more accurately, in this way the agent obtains a partial description of his desirability matrix: he knows that it is one of the numerical matrices that can be obtained by substituting definite numbers for "x," "y," and "z" throughout the foregoing generic matrix. Suppose that his probability matrix is as follows:

.25	.5	.25
.25	.5	.25

This matrix corresponds to the judgment that the probability of rain is the same (.5) on each of the two days, and is the same whether or not it rains on the other day, and whether the agent performs the first or the second act. The desirability of the first act is then

$$(.25)(x - 3) + (.5)(y - 3) + (.25)(z - 3)$$

or

$$.25x + .5y + .25z - 3 ,$$

and the desirability of the second act is

$$(.25)(x) + (.5)(y - 2) + (.25)(z - 4)$$

or

$$.25x + .5y + .25z - 2 .$$

Thus, once the numerical desirabilities x, y, z of 0, 1, and 2 days' swimming are known, the desirabilities of the two acts are known numerically. However, in order to determine which is the better act, it is not necessary to know the values of x, y, and z: whatever those numbers may be, the desirability of the second act will exceed that of the first by 1, and therefore the second act must be the better. Buying tickets daily is identifiable as the better act according to Bayesian principles even though the desirability matrix is incompletely known. In the following chapter we shall see why this sort of thing happens, and, in general, how accurately the desirability and probability matrices must be specified in order for Bayesian deliberation to have a conclusive outcome.

1.5 Dominance, and a Fallacy

In the nuclear disarmament example, monetary considerations are of no help in providing numerical desirabilities, and it is extremely difficult to settle on a set of numerical probabilities. It is sometimes argued that nevertheless, deliberation can be conclusive—that in fact, disarmament is identifiable as the better act regardless of what the probabilities may be, provided only that in the desirability matrix

d_1	d_2
e_1	e_2

each e is greater than the corresponding d:

$$e_1 > d_1 , e_2 > d_2 .$$

In terms of consequences, this proviso means that continuation of life under abhorrent conditions is preferred to extinction of life ($e_1 > d_1$), and that a golden age is preferred to continuation of life under present conditions ($e_2 > d_2$). The argument uses the fact that with this proviso, the second act dominates the first in the sense that (looking at the first column of the matrix) the second act is preferable to the first in case of war and (looking at the second column) in case of peace. Then, come what may, the second act is more desirable than the first and should therefore be performed.

But whether this conclusion is true or false, the argument is fallacious. A believer in the deterrent effect of nuclear armament might use the same desirability matrix to argue to the opposite conclusion, as follows.

Example 9: Deterrence

The believer in deterrence might agree that the desirability matrix is (say)

−100	0
−50	50

However, he would contend that disarming increases the probability of war to such an extent as to make arming the better act. In particular, he might take the probability of war to be .8 in case of disarmament, but only .1 in case of armament. Then his probability matrix would be

.1	.9
.8	.2

and accordingly, he calculates that the desirability of arming is

$$(.1)(-100) + (.9)(0) = -10$$

while the desirability of disarming is only

$$(.8)(-50) + (.2)(50) = -30 .$$

Thus he concludes that arming is the better act.

The assumption that the dominant act is the better is correct if an extra premise is introduced, i.e., that *the probabilities of the conditions are the same, no matter which act is performed.*

Example 10: Unpacking the argument for disarmament

When conditions are independent of acts the two rows of the probability matrix will be the same. Therefore, the matrix will have the form

p	$1-p$
p	$1-p$

Then the estimated desirability of the first act will be

$$pd_1 + (1-p)d_2$$

and that of the second act will be

$$pe_1 + (1-p)e_2 .$$

Since the *e*'s are greater than the corresponding *d*'s, the second expression must be greater than the first, and the expected desirability of the second act is the greater. Thus, if the particular desirabilities are

−100	0
−50	50

as before, the two expressions become

$$(p)(-100) + (1-p)(0) = -100p$$

$$(p)(-50) + (1-p)(50) = -100p + 50 ,$$

and the second expression must be greater by 50 than the first, regardless of p.

The advocate of nuclear disarmament might amplify his argument by inserting an extra premise, namely, that the probability of war, whatever it is, is practically the same no matter which act is performed. Thus, he might concede the existence of a deterrent effect of nuclear armament, but believe that this is almost exactly cancelled by an increased danger of accidental war. He may amend his argument in this way, but he cannot then present it as an argument which makes no assumptions about the probability matrix. In its original form, the argument was indeed invalid.

1.6 Problems

1 Train or Plane?

As far as cost and safety are concerned, train and plane provide equally good ways of getting from Las Pulgas to San Francisco. However, the trip takes 8 hours by train, and 3 hours by plane, unless the San Francisco airport proves to be fogged in, in which case the plane trip takes 15 hours. According to the weather forecast, there are 7 chances out of 10 that San Francisco will be fogged in. Draw up a probability matrix based on the weather forecast, and a desirability matrix based on the travel times (putting minus signs before the times given above, since short trips are preferred), and decide whether to take the train or plane.

2 The Point of Balance

In problem 1, what must the probability of fog be, if plane and train are to be equally good choices?

3 Beyond Matrices

Many problems are simplified if we use different sets of relevant conditions for different acts, in a way that is impossible in the usual matrix formulation. Example: the variant of problem 1 in which the conditions relevant to going by plane are still fog or not (at the airport), but the conditions relevant to going by train are quite different: snow or not (in a mountain pass). For the act of going by plane, the probability × desirability entries would be as in problem 1:

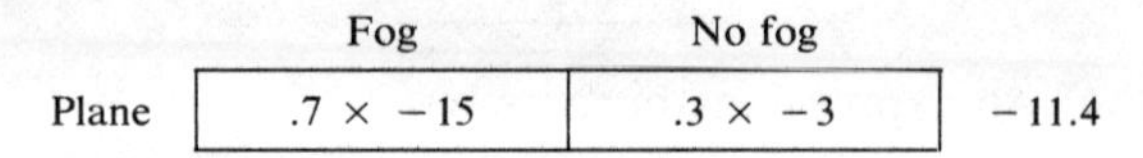

	Fog	No fog	
Plane	$.7 \times -15$	$.3 \times -3$	-11.4

(The desirability of going by plane is shown at the right.) But if the probability of snow in the pass is .5, and the train trip takes 10 hours or

8, depending on whether or not there is snow in the pass, the relevant information about going by train would be shown in a row of a different matrix, with different column headings:

	Snow	No snow	
Train	$.5 \times -10$	$.5 \times -8$	-9.0

Then, in order to see that going by train is preferable, there is no need to use the standard matrix format, as below. Problem: discuss (and, if possible, overcome) the difficulties of filling in the blanks of the standard matrix with probabilities and utilities on the basis of the information given so far.

	Fog and snow	Fog and no snow	No fog but snow	No fog and no snow
Plane				
Train				

4 The Heavy Smoker

In *The Consumers Union Report on Smoking and the Public Interest* (Consumers Union, Mt. Vernon, N.Y., 1963, p. 69), figures provided by the American Cancer Society are reproduced as follows.

Percentage of American Men Aged 35 Expected to Die before Age 65

Nonsmokers	23%
Cigar and pipe smokers	25%
Cigarette smokers:	
Less than 1/2 pack a day	27%
1/2 to 1 pack a day	34%
1 to 2 packs a day	38%
2 or more packs a day	41%

The agent is a 35-year-old American man who has found that if he smokes cigarettes at all, he smokes 2 or more packs a day. He sees his options as these three:

C = Continue to smoke 2 or more packs of cigarettes a day
S = Switch from cigarettes to pipes and cigars
Q = Quit smoking altogether

He sees the relevant conditions as these two:

D = Die before the age of 65
L = Live to age 65 or more

Supposing that his views about probabilities are derived in the obvious way from the foregoing statistics and that his notions of desirability are given by the following matrix, compute the desirabilities of the three options and identify the one he chooses if he follows the Bayesian rule.

	D	L
C	0	100
S	−1	99
Q	−5	95

5 Nicotine Addiction

Suppose that in problem 4 the agent does not see Q as an option: he lacks the willpower required to quit. Suppose, too, that his probability matrix derives from the Consumers Union figures, but that concerning his desirability matrix we know only that it has the following form, with lowest desirability (d) assigned to dying before age 65 in spite of having switched, and with l and c being positive, independent increments in desirability contributed by longevity (l) and cigarette-smoking (c).

	D	L
C	$d + c$	$d + c + l$
S	d	$d + l$

The agent does switch to pipes and cigars. Show that therefore c can have been no more than 16% of l.

6 The Dominance Principle

According to this principle (which is of limited applicability, according to the version of Bayesianism propounded here), an act is preferable to any other act that it dominates. Compare and contrast the ordering of the three acts in problem 4 by dominance with that by estimated desirability.

7 Pascal's Wager

The agent is trying to decide whether or not to undertake a course of action designed to overcome his intellectual scruples and lead to belief in God. He takes the consequence matrix to be the following.

	God exists	There is no God
Succeed in believing	Eternal life	Finite life, deluded
Remain an atheist	A bad situation	The presumed status

He takes the desirabilities to be as follows, where x and y are finite, and

z is either finite or $-\infty$, and he takes the chance of God's existing to be one in a million. What should he do?

∞	x
z	y

8 Fermat's Wager

The terms "Fermat's last theorem" and "the Fermat conjecture" are applied to the following assertion (for which Pierre de Fermat claimed to have found a marvellous proof, which was unfortunately too long to fit in the margin where the assertion itself was found, in Fermat's handwriting, after his death).

If x, y, z, and n are positive integers and
$x^n + y^n = z^n$, then n is 1 or 2 .

After three centuries, no proof or counterexample has yet been produced. Now suppose (quite implausibly) that Fermat knew he had no proof, but simply wanted to enhance his posthumous reputation as a mathematician. His problem, then, was to choose one of three acts: (*a*) assert the conjecture, (*b*) deny it, (*c*) leave the margin blank. If Fermat was quite confident that in time a proof or a counterexample would be found, the relevant conditions might be (*d*) that the conjecture is true and (*e*) that it is false. The probability matrix must have had the form shown at the left, and it might be plausible (why?) to suppose that the desirability matrix had the form shown at the right.

	d	e
a	p	$1 - p$
b	p	$1 - p$
c	p	$1 - p$

	d	e
a	$1/p$	-1
b	-1	$1/(1 - p)$
c	0	0

Knowing that Fermat performed act (*a*), what can we conclude about the probability (p) that he attributed to his conjecture?

9 The St. Petersburg Game

A coin is tossed repeatedly until the tail first turns up (on toss number n, say), at which point the player receives 2^n ducats. What is the actuarial value of this game? (The probability is $1/2^n$ that the tail first turns up on toss number n.)

10 Allais's Gambles

Which of these gambles (costless to you!) would you choose?

A = \$500,000 with 100% probability
B = \$2,500,000, \$500,000, or \$0 with probabilities 10%, 89%, 1%.

Which of these would you choose?

C = \$500,000 or \$0 with probabilities 11%, 89%.
D = \$2,500,000 or \$0 with probabilities 10%, 90%.

Note your answers, and then compute the actuarial values of the four gambles. Many people report preferring *A* to *B*, and *D* to *C*. Show that these preferences are radically un-Bayesian, in the sense that there are no numbers *x, y, z* that could represent the desirabilities someone attributes to gaining \$0, \$500,000, and \$2,500,000 so as to give *A* a higher estimated desirability than *B* while giving *C* less than *D*.

11 The Sure Thing Principle

This concerns pairs of pairs of acts, e.g., the pairs (*A, B*) and (*C, D*) of which the members are the acts of accepting free tickets in the four 100-ticket lotteries described by the following consequence matrix (in which the numbers are payoffs in units of \$100,000, so that the lotteries are realizations of Allais's four gambles).

Number of the Winning Ticket

	1	2–11	12–100
A	5	5	5
B	0	25	5
C	5	5	0
D	0	25	0
	E		*F*

The possible outcomes of the lottery can be divided into two classes, *E* and *F*, depending on whether or not the winning ticket is one of those numbered 1 through 11. Possibilities in *F* are irrelevant to preference within each pair, for if the winning ticket is numbered 12 or higher, *A* and *B* have the same payoff (5), and so do *C* and *D* (0). Then your preference (if any) between *A* and *B* should be the same as that between *C* and *D*, respectively, for *A* and *C* have the same consequences when the winning ticket is one of the first 11, and so do *B* and *D*.

(a) Formulate the sure thing principle in a general way.
(b) By means of an example, show that on Bayesian principles as propounded here, the sure thing principle is not always trustworthy. (Suggestion: consider a case where lotteries B and D are rigged, in different ways.)
(c) Under what general conditions will the sure thing principle be trustworthy by Bayesian lights?

12 The Prisoners' Dilemma

Two men are arrested for bank robbery. Convinced that both are guilty, but lacking enough evidence to convict either, the police put the following proposition to the men and then separate them. If one confesses but the other does not, the first will go free (amply protected from reprisal) while the other receives the maximum sentence of ten years; if both confess, both will receive lighter sentences of four years; and if neither confesses, both will be imprisoned for one year on trumped-up charges of jaywalking, vagrancy, and resisting arrest. Draw up a desirability matrix for one of the prisoners and explain why the police are convinced that both will confess, even though both would be better off if neither confessed.

13 Egoism Doesn't Always Pay

Suppose that the prisoners in problem 12 are revolutionaries who care for nothing but their cause, and who regard each other as equally valuable to it. Draw up a desirability matrix based on total man-years in prison, and explain why neither prisoner will confess.

14 Newcomb's Problem

A preternaturally good predictor of human choices does or does not deposit a million dollars in your bank account, depending on whether he predicts that you will reject or accept the extra thousand that he will offer you just before the bank reopens after the weekend. Would it be wise on Monday morning to decline the bonus?

15 Fisher's Problem

Suppose that the nicotine addict in problem 5 is convinced that the statistical association between smoking and cancer recorded in problem 4 does not implicate smoking as a cause of cancer, but is due to the fact that smoking and cancer are independent, probable results of a common cause, i.e., a certain genetic makeup. (If you have the bad gene, you are more likely to get cancer than if you haven't, and you are more likely to find that smoking dominates abstaining. But if you have the bad gene, smoking will not make cancer more likely than abstention would; nor will it if you do not have the bad gene. That's what the agent believes.) What should the agent's probability matrix look like? What should he do?

1.7 Ratifiability

Any decision must be expected to change some of the agent's subjective probabilities. For example, in the prisoners' dilemma (problem 12), when a prisoner decides to confess, the probability he attributes to his actually confessing gets close to 1. (It should not go all the way to 1, for he ought to know that some misadventure might prevent his actually carrying out his decision.) In ordinary problems, such changes in the probabilities of actions do not change the matrix of probabilities of conditions, given the various actions. But in the bothersome sorts of cases exemplified by the prisoners' dilemma, the probability matrix can be expected to change in one way or another as and when one act or another is chosen. The question of whether or not a particular course of action has highest estimated desirability can then get different answers, depending on whether the estimates are made with the probability matrix that reflects the deliberating agent's uncertainty about how he will finally choose, or with the matrix that would replace that one upon choice of a particular one of the options.

Definition. A ratifiable decision is a decision to perform an act of maximum estimated desirability relative to the probability matrix the agent thinks he would have if he finally decided to perform that act. (This will be explained and illustrated as we go.)

Maxim. Make ratifiable decisions. To put it romantically: "Choose for the person you expect to be when you have chosen."

Example 11: The prisoners' dilemma

The police correctly expect both prisoners to choose the dominant act: confess. But the Bayesian principle advises each prisoner *not* to confess, if each sensibly sees his own choice as a strong clue to the other's and therefore assigns high subjective probabilities (p, q) to the other prisoner's doing whatever it is (confess, do not confess) that he himself chooses: see table 1.1, where the estimated desirability of choosing not to confess exceeds that of choosing to confess if p and q are 77% or more.

Table 1.1

	He does confess	He does not	
I choose to confess	$p \times -4$	$(1 - p) \times 0$	$-4p$
I choose not to	$(1 - q) \times -10$	$q \times -1$	$9q - 10$

Deciding to confess has the lower estimated desirability if p and q (the probabilities I attribute to my decision's being an accurate indicator of his performance) both exceed 10/13.

Now if I am nearly certain that my performance will accurately indicate my final choice, e.g., if the probabilities are these,

	I chose to confess	I chose not to
I shall confess·	.95	.05
I shall not	.03	.97

then p and q as in table 1.1 will closely approximate the probabilities I assign to the other prisoner's doing whatever it is that I *do*, and so the estimates ($-4p$, $9q - 10$) given in table 1.1 for the desirabilities of my two *choices* will be nearly accurate as estimates of the desirabilities of my two *performances*. Thus, on plausible assumptions, "Don't confess" will be the Bayesian advice, relative to the probability matrix that reflects my uncertainty about how I shall finally choose. But when desirabilities of acts are estimated in the light of anticipated post-choice probabilities, as in table 1.2, the dominant act (confess) is seen to have the higher estimated desirability on each hypothesis about my final decision. Table 1.2 takes account of the fact that on each hypothesis about his decision, each prisoner ("I") sees his own performance as predictively irrelevant to the other's—and sees that as he deliberates. So it is that in part (a) of that table, and also in part (b), probabilities are the same in both rows.

Table 1.2.
Estimated Desirabilities of Actual Performances if

(a) I decide to confess

	He does	doesn't	
I do	$p \times -4$	$(1-p) \times 0$	$-4p$
don't	$p \times -10$	$(1-p) \times -1$	$-9p-1$

(b) I decide not to confess

	He does	doesn't	
I do	$(1-q) \times -4$	$q \times 0$	$4q-4$
don't	$(1-q) \times -10$	$q \times -1$	$9q-10$

As probabilities are independent of acts in each matrix, my dominant performance (confess) will have the higher estimated desirability on each hypothesis about my final decision.

The prisoners' dilemma exemplifies a class of problems in which the agent foresees during deliberation that choice will change subjective probabilities so as to make the dominant act be the one whose performance will have highest estimated desirability once a decision (either decision) has been made, even though it did not have highest estimated desirability when the agent's uncertainty about his final decision was taken into account, before choice. It seems clear that one should choose an act whose performance one expects to have highest estimated desirability if and when it is chosen. To put the same thing the other way around, I should not make a decision I would wish not to carry out, e.g., in the prisoners' dilemma, the decision not to confess. (Table 1.2(b) shows me in advance that I would rue that choice as soon as I had made it, for no matter what value q may have, from 0 to 1, the estimated desirability $4q - 4$ of confessing in spite of having decided not to exceeds the estimated desirability $9q - 10$ of actually carrying out a decision not to confess.)

"Choose an act of maximum estimated desirability." That is the Bayesian maxim as formulated at the beginning of this chapter. But in the prisoners' dilemma, and in other, more far-fetched cases such as Newcomb's problem, that maxim produces bad advice: "Don't confess" and "Don't take the extra thousand." Ratificationism modifies the Bayesian maxim in a way that makes no difference in the usual, straightforward sorts of problems, but does make the right sort of difference in problems where the agent takes his decisions to have strong evidentiary significance concerning conditions that he believes his performances have no strong tendency to promote or prevent. Ratificationism requires performance of the chosen act, A, to have at least as high an estimated desirability as any of the alternative performances *on the hypothesis that one's final decision will be to perform A*.

The notion of ratifiability is applicable only where, during deliberation, the agent finds it conceivable that he will not manage to perform the act he finally decides to perform, but will find himself performing one of the other available acts instead. In the prisoners' dilemma, death or a nonfatal cerebral hemorrhage might prevent me from carrying out a decision to confess, and a surreptitiously administered dose of sodium pentothal might set me to confessing in spite of my decision not to. But changing my decision does not count as a way of not managing to perform my chosen action. In table 1.2, the hypotheses (a) and (b) of the reasoning that identifies confession as ratifiable and nonconfession as unratifiable are to be understood as hypotheses about my final decision. No matter that I may not know at the time that my current decision is final; it is enough that as part of my deliberation I can entertain the various hypotheses about what decision I shall finally make.

In exceptional cases, none of the available acts is ratifiable (example 12), or all are (example 13).

Example 12: The green-eyed monster

Where the agent must choose one of two goods and see the other go to someone else, greed and envy may conspire to make him rue either choice. Decision theory cannot cure this condition, and ratificationism recommends neither option.

Example 13: The triumph of the will

A madly complacent agent could find all acts ratifiable because with him, choice of an act always greatly magnifies his estimate of its desirability—not by changing probabilities of conditions, but by adding a large increment to each entry in the chosen act's row of the desirability matrix. And any of us will sometimes see both horns of a dilemma of acts as ratifiable, either because of such a desirability increment or because of the anticipated effect of choice on the probabilities of conditions.

Ratificationism, what. Ratificationism is the doctrine that the choiceworthy acts—relative to your beliefs and desires—are the ratifiable ones. Pathology is to be suspected when the number of ratifiable options is not exactly 1: in such cases you do well to reassess your beliefs and desires before choosing. On the other hand, prisoners' dilemmas are cases of cruel circumstances, not agent pathology.

Where the probability matrix is the same on each hypothesis about which act will finally be chosen, ratificationism does not conflict with the Bayesian principle as formulated at the beginning of this chapter. That sort of independence of conditions and choices is the norm when (as in problems 4 and 5) the agent sees the differences between rows of his probability matrix only as reflections of the tendencies of different acts to promote or prevent different conditions. The prisoners' dilemma is odd in that there, the agent sees his choices as ineffectual harbingers of conditions that he would promote or prevent if he could.

What ratificationism is not. It is not the modification of the Bayesian principle in which the desirability of each act is estimated as it would be if that very act had just been chosen, and comparative choiceworthiness goes by desirability so estimated. On that reckoning, the desirabilities of confessing and not confessing would be estimated as $-4p$ and $9q - 10$ (being derived from different matrices in table 1.2), and the result would be no different from that obtained by using table 1.1: don't confess, if you are pretty sure that he is much like you. But on the contrary, we are to judge the ratifiability of each act by estimating the desirabilities of all acts from the point of view the agent would have upon choosing the one act whose ratifiability is in question. Ratifiability is a classificatory notion, not a comparative one.

How to determine whether a particular option is ratifiable. On the hypothesis that *that* option will finally be chosen, estimate the desirabilities of actually carrying it out, and of actually carrying out each of the alternatives. The option in question is ratifiable or not depending on whether or not the expected desirability of actually carrying it out (having chosen it) is at least as great as the expected desirability of actually carrying out each of the alternatives (in spite of having chosen to carry out a different option, as hypothesized).

In a nutshell: suppose that among the performances you might choose, none has higher estimated desirability on the hypothesis that A is chosen than A itself has. Then, and only then, is A ratifiable.

It would be pleasant to be able to report that choiceworthiness always corresponds to ratifiability, as it did in the prisoners' dilemma of example 11. But that cannot be, for in the following more plausible variant of that example (suggested by Bas van Fraassen) confessing remains choiceworthy but need not be ratifiable.

Example 14: The prisoners' dilemma II

In table 1.2(a) the probability, p, of his confessing in case I do (having decided to) is no greater than the probability, p', of his confessing in case I don't (still having decided to). Indeed we have $p' = p$ there, so that in the lower row of matrix (a) "p" could appear with no accent. But in a more plausible telling of the story, p' is smaller than p because the sorts of extraneous influences that would prevent me from confessing when I had decided to are likely to work on him, too, and to the same effect. Then let us put accents on the three occurrences of "p" in the second row the the (a) matrix. This means that, when I have decided to confess, the desirabilities of my confessing and not confessing will be $-4p$ and $-9p' - 1$. Confessing will be unratifiable in case the first of these is the lesser, as it may well be. Thus, in case $p = 3/4$ and $p' = 1/9$ we have $-4p = 3$, which is less than $-9p' - 1 = -2$. The same effect can be produced with a smaller discrepancy between p and p' if third best is pushed closer to worst, e.g., if the story is told with a 9-year term in place of a 4-year term confessing will be unratifiable as long as p exceeds p' by more than 1/9.

Conclusion. Ratibiability is not a completely dependable guide to choiceworthiness. While the maxim "Make ratifiable decisions" is dependable in a broader range of cases than the old maxim, i.e., "Choose maximally desirable acts," there remain cases where the new maxim may mislead. These are cases like example 14, where actions as well as choices portend states of nature that they have little or no tendency to bring into being.

1.8 Notes and References

The epigraph is from the final chapter of the Port-Royal Logic, i.e., *La logique, ou l'art de penser* (Paris, 1662), by Antoine Arnauld. An

English translation is available: *The Art of Thinking* (Indianapolis: Bobbs-Merrill, Library of Liberal Arts, 1964).

Bayes's work was published shortly after his death, a century after publication of the Port-Royal Logic: "An Essay toward Solving a Problem in the Doctrine of Chances," *Philosophical Transactions of the Royal Society of London,* 1763. It has been reprinted in *Biometrika* 45 (1958): 293–315, and in *Studies in the History of Statistics and Probability,* ed. E. S. Pearson and M. G. Kendall (London: Charles Griffin, 1970).

Pascal's wager was the centerpiece of his posthumously assembled *Pensées.* It is numbered 233 in many arrangements, and 418 in recent ones, e.g., in A. J. Krailsheimer's translation (Penguin Books, 1966). It first saw light in the Port-Royal Logic, where it occupies the place of honor, at the very end.

What I call "estimated desirabilities" would be called "conditional expected utilities" in the usual technical jargon. "Utility" has misleading philosophical associations—notably, with Bentham's hedonism—but "desirability" misleads if it suggests what is really desirable in contrast to what the agent does in fact desire. The "-ability" ending matches that in "probability," which here refers to the agent's credence as it is, not as it should be.

The version of Bayesianism floated here was a novelty in 1965, when this book first appeared, because it did not require all rows of the agent's probability matrix to be the same, i.e., it allowed the relevant conditions to have different probabilities depending on which act is performed. The innovation is natural and useful, e.g., allowing us to use a two-column matrix in problem 4, where insistence that probabilities of conditions be the same for all acts would force us to resort to some such artifice as that suggested by Leonard J. Savage in *The Foundations of Statistics* (New York: Wiley, 1954; 2d ed., revised, New York: Dover, 1972), p. 15. For a difficult, full account of such an artifice, see David H. Krantz, R. Duncan Luce, Patrick Suppes, and Amos Tversky, *Foundations of Measurement,* vol. 1 (New York: Academic Press, 1971), pp. 414–16. An easier, sketchier account (in my paper "Savage's Omelet," *PSA 1976,* vol. 2 [East Lansing, Michigan: Philosophy of Science Association, 1977], pp. 361–71) is illustrated in example 15.

Example 15: Savage's treatment of the heavy smoker (problem 4)
In place of our 3-by-2 probability and desirability matrices

	D	*L*
C	.41	.59
S	.25	.75
Q	.23	.77

	D	*L*
C	0	100
S	−1	99
Q	−5	95

C	59
S	74
Q	72

yielding the desirability estimates shown in the column at the right, Savage would use a 3-by-8 combined probability and desirability matrix that yields the same desirability estimates:

	.0236 *CSQ* *DDD*	.0789 *CSQ* *DDL*	.0707 *CSQ* *DLD*	.2368 *CSQ* *DLL*	.0339 *CSQ* *LDD*	.1136 *CSQ* *LDL*	.1018 *CSQ* *LLD*	.3407 *CSQ* *LLL*	
C	0	0	0	0	100	100	100	100	59
S	−1	−1	99	99	−1	−1	99	99	74
Q	−5	95	−5	95	−5	95	−5	95	72

The probabilities of the eight relevant conditions (the same for each act) are listed above them, to four decimal places. The column headings specify the eight possible hypotheses about the consequences of the three acts: *CSQ/XYZ* specifies that condition *X* or *Y* or *Z* will hold, depending on whether act *C* or *S* or *Q* is performed. To that hypothesis we assign, as its act-independent probability, the product *xyz* of the entries in the *C* row and *X* column, *S* row and *Y* column, and *Q* row and *Z* column of the 3-by-2 probability matrix. Thus, *CSQ/DDD* says that the agent will die before age 65 no matter what he does, and its probability is determined by $x = .41$, $y = .25$, $z = .23$.

Savage had his reasons for insisting that conditions (he said "states" or "states of nature") be probabilistically independent of acts. But his reasons lose cogency when (as here) we depart from the special framework that he adopted—a framework in which acts are taken to assign definite consequences to states, so that to know what act you are performing you must know exactly how it would turn out in each possible state of nature. Then the probability of any particular consequence, given an act and a state, can only be 1 or 0, depending on whether that consequence is or is not the definite outcome of that act in that state. In such a scheme, the only chink through which ordinary probabilities can be introduced is at the top, as in the 3-by-8 matrix above. For more on this topic, see my "Frameworks for Preference," in *Essays on Economic Behavior under Uncertainty*, ed. Michael Balch, D. McFadden, and S. Wu, (New York: Elsevier, 1974).

The St. Petersburg game was first discussed by Daniel Bernoulli, "Specimen theoriae novae de mensura sortis" [Exposition of a new theory of the measurement of risk], in the *Proceedings of the St. Petersburg Imperial Academy of Sciences*, vol. 5, 1738. A translation by Louise Sommer appears in *Econometrica* 22 (1954): 123–36, and has been reprinted in *Utility Theory: A Book of Readings*, ed. Alfred N. Page (New York: Wiley, 1968).

Maurice Allais's gambles were first discussed in "Le comportment de l'homme rationnel devant la risque: critique des postulats et axiomes de l'école Américaine," *Econometrica* 21 (1953): 503–46. For a clear discussion in English, see Savage, *Foundations of Statistics,* pp. 101–3.

The sure-thing principle was so called by Savage (pp. 21–23), who adopted it as a postulate, i.e., one of those that Allais criticises. Like the dominance principle it becomes valid when restricted to problems in which all rows of the probability matrix are the same.

Allais allowed that in problem 12, A might be preferable to *B* while *D* was preferable to *C*—a possibility excluded by even the present, liberal version of Bayesianism.

The prisoners' dilemma is A. W. Tucker's two-person zero-sum "game," beloved of social scientists, about which a large body of literature has grown in the past twenty-five years. As David Lewis points out in "Prisoners' Dilemma Is a Newcomb Problem" (*Philosophy and Public Affairs* 8 [1979]: 235–40), it is equivalent to what is perhaps the most interesting reading of a problem invented by William Newcomb that has been much discussed by philosophers in the past dozen years. (Problem 14 is a particularly clear version of it devised by Howard Sobel.) The problem was aired, and named, by Robert Nozick ("Newcomb's Problem and Two Principles of Choice," in *Essays in Honor of Carl G. Hempel,* ed. Nicholas Rescher [Dordrecht: D. Reidel, 1969]). It has subsequently been seen as a rock on which the sort of Bayesianism floated early in this chapter must founder. I survey some of that literature in "*The Logic of Decision* Defended," *Synthese* 48 (1981): 473–92 (where there is a typographical omission: add the words "deliberate by reasoning hypothetically about our credences and" at the bottom of page 486). What is mounted there as a defense reappears more aptly here in section 1.7 as a *modification* of the form of the Bayesian maxim given at the beginning of this chapter. The defense/modification was inspired by Ellery Eells's Ph.D. dissertation (Berkeley, 1980), which is excerpted in *Synthese* 48 (1981): 295–329, and further elaborated in his book, *Rational Decision and Causality* (Cambridge: At the University Press, 1982). My two sorts of probabilities—hypothetical on final choices, and otherwise—were inspired by Eells's type *B* and type *A* beliefs, but whereas Eells's contrast is between the agent's beliefs about his own predicament and about the predicaments of all people in his sort of trouble, mine is between the agent's conditional and unconditional beliefs about his own predicament. Although inspired by it, ratificationism differs from Eells's approach, and must be judged independently.

Nozick posed a problem without offering a general solution. Ratificationism aims to be a general solution, in which choiceworthiness is no longer defined as a matter of being at the top of the ranking of options according to estimated desirability (although it comes down to that in all but Newcomb-like cases). It fails in the cases noted at the end of section 1.7. Another family of solutions is surveyed and strongly promoted by David Lewis in "Causal Decision Theory," *Australasian Journal of Philosophy* 59 (1981): 5–30, where further references can be found. (See Eells [1982] chapter 5 for a survey of some of those solutions from an antagonistic point of view.) They emerged pretty much at once in the late 1970s, in the work of Alan Gibbard and William Harper, Howard Sobel, Brian Skyrms, Nancy Cartwright, and Lewis himself. The inspiration for much of that work came from Robert Stalnaker: see *Ifs*, ed. W. L. Harper, R. Stalnaker, and G. Pearce (Dordrecht: D. Reidel, 1981), pp. 151–52. The Gibbard and Harper work is reprinted there too (pp. 153–90). These solutions keep the comparative character of choiceworthiness, but break the connection between choiceworthiness and estimated desirabilities of acts. Thus, in Brian Skyrms's treatment (in part 2c of *Causal Necessity* [New Haven: Yale University Press, 1980]), the choiceworthiness of an act A is a weighted average

$$des(C_1A)prob(C_1) + \ldots + des(C_nA)prob(C_n)$$

of the estimated desirabilities $des(C_iA)$ it would have under n incompatible, exhaustive conditions that the agent takes to be outside his influence (e.g., the other prisoner's action, or the presence of the million dollars in his bank account), where the weights are the unconditional probabilities $prob(C_i)$ that he attributes to those conditions. This can differ from the estimated desirability of A

$$des(A) = des(C_1A)prob(C_1/A) + \ldots + des(C_nA)prob(C_n/A) ,$$

(where the conditional probabilities $prob(C_i/A)$ are ratios $prob(C_iA)/prob(A)$ of unconditional ones) in case the agent sees performance of A as a clue to which of the conditions holds, even though he takes his act to have no causal influence on those conditions (as in the prisoners' dilemma and Newcomb's problem), or anyway takes the evidentiary significance of his act to be greater than its causal influence accounts for. In this notation, option A is ratifiable if and only if the inequality

$$des(A \text{ when } A \text{ is chosen}) \geqslant des(B \text{ when } A \text{ is chosen})$$

holds for every option, B.

For an exploration of the ambiguities of Newcomb's problem, see Isaac Levi, "Newcomb's Many Problems," *Theory and Decision* 6 (1975): 161–75. For a view on which the Bayesian principle is satisfactory as formulated at the beginning of this chapter (and on which one should not take the extra thousand in Newcomb's problem, and, presumably, should not confess in the prisoners' dilemma), see Maya Bar-Hillel and Avishai Margalit, "Newcomb's Problem Revisited," *British Journal for the Philosophy of Science* 23 (1972): 295–304. For a medley of responses to Newcomb's problem, see Robert Nozick's "Reflections on Newcomb's Problem: A Prediction and Free Will Dilemma," *Scientific American* 230 (1974): 102–8. With David Lewis, I see Newcomb's problem as a prisoners' dilemma for space cadets: a secular, sci-fi successor to the problems of predestination that exercised such thinkers as Jonathan Edwards (1703–58). It is a problem to which I (unlike Lewis) would apply the thought of Esther Marcovitz (1866–1944): "If cows had wings, we'd carry big umbrellas." Our prisoners were paradox enow.

Problem 15 was inspired by R. A. Fisher's pamphlet *Smoking: The Cancer Controversy* (Edinburgh and London: Oliver and Boyd, 1959). Observe that if it is the *desire* to smoke that the agent takes to be directly promoted by the bad gene (so that the choice and the performance are promoted only indirectly), the agent's awareness of the tickle of desire (i.e., of the fact that for him, smoking dominates abstention) would lead him to a probability matrix with identical rows, so that unreconstructed Bayesianism would endorse the dominant act. Fisher's problem belongs to the Newcomb species only if it is the *choice* of smoking that the agent takes the bad gene to promote directly. If he takes the desire to be directly promoted, probabilities of conditions will be unaffected by choice. If he takes the performance itself to be directly promoted by presence of the bad gene, there is no question of preferential choice: the performance is compulsive.

2

Equivalent Scales

In the straightforward sort of Bayesian deliberation in which choice does not change the agent's estimates of desirabilities, the point of computing such estimates is to establish a preference ranking of the available acts. The order of acts in the preference ranking is identified with the numerical order of their estimated desirabilities, and the Bayesian recommendation is that one of the top-ranking acts (or *the* top-ranking act, if there is only one) be performed.

Example 1

Suppose that in example 3 of chapter 1, the guest might bring a bottle of white, red, or rosé wine, and that the main dish might be chicken, beef, or herring. The guest's desirability matrix might then be this:

	Chicken	Beef	Herring
White	1	−1	1
Red	0	1	−1
Rosé	.5	0	−1

If he takes the probabilities of chicken, beef, and herring to be .4, .4, and .2, regardless of which act he performs, his probability matrix will be as follows (left-hand columns):

White	.4	.4	.2	.2
Red	.4	.4	.2	.2
Rosé	.4	.4	.2	0

so that the expected desirabilities of bringing white, red, and rosé are as in the column at the right. Then with the given probability and desirability matrices, the preference ranking of acts is this:

White, Red

Rosé

Here the preferred options appear higher on the list, and options shown on the same line are equally good. Had the probability matrix been different, a different preference ranking of the acts might have ensued. Thus, with the same desirability matrix but with probabilities as at the left,

.4	.2	.4
.3	.4	.3
.4	.4	.2

.6
.1
0

expressing the guest's belief in a certain influence of his choice of wine on the hosts' choice of food, the expected desirabilities would be as shown at the right yielding a preference ranking

White

Red

Rosé

in which the tie between white and red is broken.

2.1 Equivalent Desirability Matrices

A single desirability matrix, used with different probability matrices, may yield different preference rankings of the possible acts. (This fact figured large in our treatment of the prisoners' dilemma in section 1.7; but now we are confining attention to the straightforward cases, where choice does not change the probability matrix.) Clearly, the reverse situation is also possible: different desirability matrices, used with a common probability matrix, may well yield different preference rankings of the acts. Whether the resulting preference rankings are the same or different in a particular case may depend on the particular probability matrix that is used in estimating desirabilities. But if the two desirability matrices are related in certain special ways that we shall now study, they can be depended upon to yield a common preference ranking of acts no matter what common probability matrix is used. In such cases the two desirability matrices are said to be *equivalent*.

Example 2

The following three desirability matrices are equivalent. (The rows correspond to the acts of bringing the white, red, or rosé as in example 1.)

1	−1	1
0	1	−1
.5	0	−1

2	−2	2
0	2	−2
1	0	−2

1	−3	1
−1	1	−3
0	−1	−3

The first matrix is identical with the desirability matrix of example 1. The second is obtained from the first by doubling each entry. The third is obtained from the second by diminishing each entry by 1. Later, we shall see that the equivalence of these three matrices illustrates a general rule: that when each entry in a matrix is multiplied by the same positive number, the result is always an equivalent matrix, and when each entry is augmented by the same amount, or diminished by the same amount, the result is always an equivalent matrix. But before proving this, let us recall the meaning of the claim that the three matrices above are equivalent. The claim means that, relative to any one probability matrix, all three desirability matrices yield one and the same preference ordering of the three acts. Thus, relative to the probability matrix at the left below, the three desirability matrices yield the three different columns of estimated desirabilities shown at the right:

.4	.4	.2
.4	.4	.2
.4	.4	.2

.2
.2
0

.4
.4
0

−.6
−.6
−1

But the three columns of estimated desirabilities determine one and the same preference ordering of acts:

White, Red

Rosé

Similarly, with the common probability matrix shown at the left here,

.4	.2	.4
.3	.4	.3
.4	.4	.2

.6
.1
0

1.2
.2
0

.2
−.8
−1

the three desirability matrices yield the three columns of estimated desirabilities of acts shown at the right—all three of which determine the same preference ordering of acts:

White

Red

Rosé

It is easy to see that whenever all entries in a desirability matrix are multiplied by the same positive number, a, the resulting matrix is equivalent to the old. (If a is 1, the resulting matrix is actually identical with the old.) For definiteness, let us assume that the matrix has three columns: the reasoning would be precisely parallel for a matrix of two, or four, or a thousand columns. Focus attention on some row of the matrix,

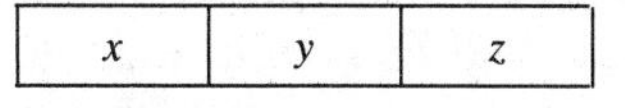

x	y	z

which is converted into the row

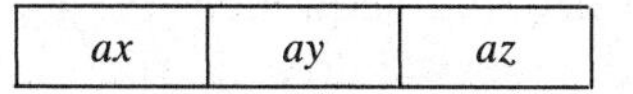

ax	ay	az

of the new matrix when all entries in the original matrix are multiplied by a. Where the corresponding row of the probability matrix is

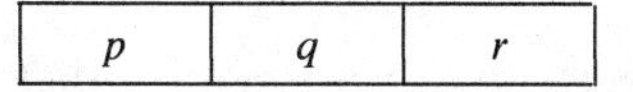

p	q	r

the expected desirability of the act in question will be

$$px + qy + rz$$

if the original desirability matrix is used, and

$$pax + qay + raz$$

if the new desirability matrix is used. Factoring the a out of each term in the second expression, we have

$$a(px + qy + rz) ,$$

which is a times the first expression. This shows that when all entries in a desirability matrix are multiplied by a number, the estimated desirabilities associated with the various rows are multiplied by the same number. Then if the estimated desirabilities are given by the column at the left (below) when the original desirability matrix is used, they will be given by the column at the right when the new desirability matrix is used:

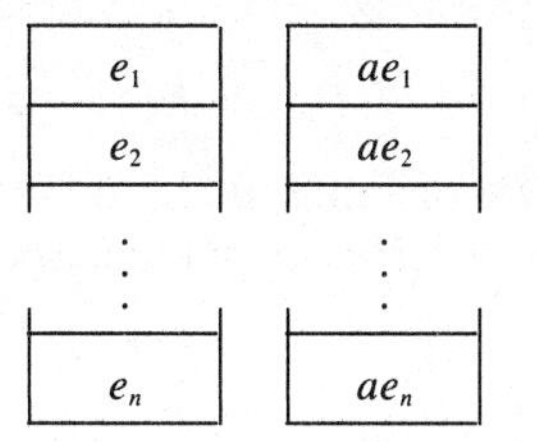

e_1		ae_1
e_2		ae_2
⋮		⋮
e_n		ae_n

Therefore, if a is positive, the numerical order of estimated desirabilities will be the same in both columns, old and new. Thus, in example 2, where the second desirability matrix came from the first by multiplying all entries by $a = 2$, the second columns of estimated desirabilities were obtainable in the same way, by doubling the three entries in the first. (Beside the first probability matrix, .2, .2, 0 in the first column became .4, .4, 0 in the second, the beside the second probability matrix, .6, .1, 0 became 1.2, .2, 0.)

In this reasoning it is essential that a be positive, because if (say) e_1 is greater than e_2, then ae_1 will be greater than, less than, or equal to ae_2 accordingly as a is positive, negative, or 0. Then, when a is negative, the numerical order of the sequence $ae_1, ae_2, \ldots, ae_n$ is just opposite to the numerical order of the original sequence, $e_1, e_2, \ldots, e_n$: the effect of multiplying each entry in a desirability matrix by the same negative number is to reverse the preference ordering (relative to a fixed probability matrix). The effect of multiplying all entries by 0 is to collapse the ordering: all acts have estimated desirability 0 and are therefore equally acceptable.

There is another sort of transformation that, applied to any desirability matrix, is guaranteed to produce an equivalent matrix, i.e., *add* the same number, b, to each entry, so that the row

x	y	z

of the original matrix is converted into the row

$x + b$	$y + b$	$z + b$

of the new matrix. The number b here may be positive, negative, or 0. Then if the corresponding row of the probability matrix is

p	q	r

the estimated desirability of the act will be

$$px + qy + rz$$

or

$$p(x + b) + q(y + b) + r(z + b) \,,$$

depending on whether the original or the new desirability matrix is used. Multiplying and collecting terms in the second of these expressions, we have

$$px + qy + rz + pb + qb + rb \,.$$

Factoring the b's we have

$$px + qy + rz + b(p + q + r) .$$

And making use of the fact (discussed in section 2.2 below) that

$$p + q + r = 1 ,$$

we see that the second expression is equal to

$$px + qy + rz + b ,$$

which is the first expression plus b. Thus the effect of adding the same number to each entry in a desirability matrix is to add that number to the estimated desirabilities of all acts. Then the estimated desirabilities of the acts will be as follows, when the old and new desirability matrices are used:

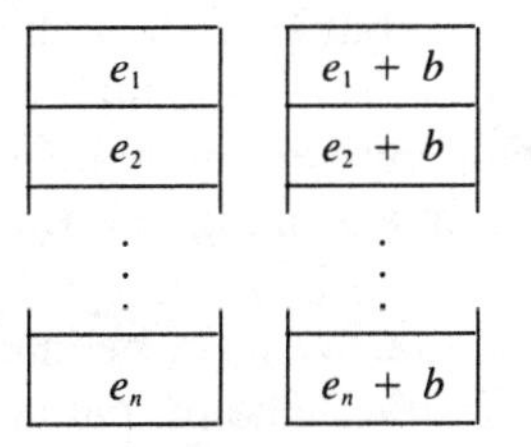

Inevitably, the numerical order of entries in the second column is the same as that in the first.

2.2 Conventions about Probabilities

Adding the same number to each entry in a desirability matrix yields an equivalent desirability matrix. In proving this, we had to make explicit use of a fact used implicitly every time we have written a probability matrix: that the entries in each row add up to 1. The reason for this is that the entries in a row of a probability matrix represent the probabilities of a set of possible conditions which are mutually exclusive and collectively exhaustive of the possibilities; and by convention, the probabilities of the conditions in such a set must sum to 1. More accurately, we should say that the entries in a row of a probability matrix represent the probabilities of the several conditions, on the assumption that the act corresponding to that row will be performed. It is this assumption that makes it possible for the rows of a probability matrix to differ from each other, as in the second probability matrix in example 1.

The convention that the probabilities of a set of mutually exclusive, collectively exhaustive conditions must sum to 1 could perfectly well be changed without requiring any compensating change in the technique of Bayesian deliberation. One such alternative convention would be that the probabilities of such a set of conditions must sum to 100; to change a probability matrix satisfying the old convention into one satisfying the new, it would be sufficient to multiply all entries by 100. With the new convention, estimated desirabilities would be 100 times as great as those which were obtained under the old convention. The estimated desirabilities $e_1, e_2, \ldots$ of a set of acts would be changed into $100e_1, 100e_2, \ldots$ Thus, the estimates .2, .2, 0 that appeared in example 1 would become 20, 20, 0; but the preference ordering of acts would not thereby be changed. In general, if we multiply each entry in a probability matrix by the same positive number, we can carry out Bayesian deliberation as before, and arrive at precisely the same conclusions: the preference ranking of the acts cannot be affected by such a transformation. The essential convention about the sums of the entries in the rows of a probability matrix is not that all such sums must be 1, but that all such sums be the same positive number. It is immaterial whether this number be 1, or 100, or 23.65; but it is useful to adopt a single convention once and for all, and, in fact, we shall take the usual course of assuming that the sum in question is always 1.

It is also worth noting that we have been using a further convention: probabilities cannot be negative. The probability of a condition is always a number in the interval from 0 to 1, with 0 representing a conclusive judgment that the condition does not obtain, with 1 representing a conclusive judgment that the condition does obtain, and with shades of doubt being represented by intermediate values. This will be discussed further below (see sections 4.2 and 5.1). For the present, it is sufficient to be clear about what the conventions are and to have a rough idea of their application.

2.3 A General Desirability Transformation

We have seen that when all the entries in a desirability matrix are multiplied by the same positive number a, an equivalent matrix results, and we have seen that when a single number b (positive, negative, or zero) is added to all entries in a desirability matrix, an equivalent matrix results. The two transformations can also be combined: if a positive number a and an arbitrary number b are chosen, and each entry

$$d$$

in a desirability matrix is changed into a new entry

$$ad + b\,,$$

the new matrix will always be equivalent to the old. It follows that in any Bayesian deliberation, we can choose the desirabilities of two differently ranked consequences at our pleasure, provided we assign the larger number to the better-liked consequence. Thus, if we ignore the trivial case in which all consequences are liked equally, we are perfectly free to assign desirability 1 to a best-liked consequence and desirability 0 to a least-liked consequence simply by replacing the original desirability matrix with an equivalent one which assigns desirabilities 1 and 0 to the two consequences. This can always be done by a suitable choice of the numbers a and b.

Example 3

In discussing nuclear disarmament (section 1.5) we used the following desirability matrix.

	War	Peace
Arm	−100	0
Disarm	−50	50

Let us now find an equivalent desirability matrix in which the extreme desirabilities are 0 and 1. The required matrix will have the form

0	?
?	1

where the two unknown numbers are to be determined so as to make the new matrix equivalent to the old. This new matrix is obtained from the old by choosing a suitable positive number a, and a suitable number b, and then replacing each entry d in the old matrix by a new entry $ad + b$. Then the original matrix becomes

$(-100)a + b$	$(0)a + b$
$(-50)a + b$	$(50)a + b$

The correct values of a and b can be found by noting that the upper left-hand entry in the new matrix is to be 0, so that

$$-100a + b = 0 ,$$

and that the lower right-hand entry is to be 1, so that

$$50a + b = 1 .$$

These equations can be solved for a by subtracting each side of the first from the corresponding side of the second. Thus, we obtain

$$150a = 1$$

or

$$a = 1/150 .$$

This value can then be substituted for a in the first equation to obtain

$$(-100)(1/150) + b = 0$$

or

$$b = 100/150 .$$

Then the required transformation replaces each entry d in the old desirability matrix by the new entry

$$(1/150)d + (100/150) .$$

In particular, it replaces the upper right-hand entry 0 by (1/150) (0) + (100/150), which is

$$2/3 ,$$

and it replaces the lower left-hand entry -50 by (1/150) (-50) + (100/150), which is

$$1/3 .$$

Thus the two unknown entries are determined in the new matrix, which is therefore known to be

0	2/3
1/3	1

The transformation of desirability matrices illustrated in example 3 is a change of scale, entirely analogous to the change which is involved when we transform a set of temperature readings from the Centigrade to the Fahrenheit scale: if the temperature of an object in degrees Centigrade is d, the temperature of the same object in degrees Fahrenheit is $ad + b$, where $a = 9/5$ and $b = 32$. We have shown that any such transformation converts a desirability matrix into an equivalent desirability matrix, and we shall say that such a transformation converts a desirability *scale* into an equivalent desirability scale.

2.4 A Special Desirability Transformation

In some cases, exemplified by the swimming example of chapter 1, we are justified in assuming that all rows of the probability matrix are identical. This will hold whenever it is clear to the agent that his performance of one act rather than another can have no influence on the possible conditions which affect the results of his actions.

Example 4

My decision whether to wash my car or not has no effect on the weather. Then, even if I have no idea of the probability of rain tomorrow, I know that my probability matrix should have the following form.

	Rain	Clear
Wash	p	$1 - p$
Don't	p	$1 - p$

My desirability matrix might be

−2	1
−1	0

which reflects my feelings about the weather, about having a clean car, and about wasting the effort required to make the car clean.

We know that when all entries in the desirability matrix are multiplied by the same positive number a, or are augmented by the same number b, the outcome of my deliberation will not be changed. But because of the special circumstance that the rows of the probability matrix are identical, an even wider class of transformations exists that can be applied to the desirability matrix of example 4 without affecting the outcome of the deliberation. I can pick a single positive number a, and two numbers b_1 and b_2; I can subject the desirabilities in the first column of the matrix to the transformation which changes d into

$$ad + b_1 ,$$

I can subject the desirabilities in the second column to the transformation which changes d into

$$ad + b_2 ,$$

and I can be sure that the resulting matrix will yield the same preference ranking of acts as the original desirability matrix, provided the same probability matrix is used in both cases, and provided this probability matrix has identical rows. Thus, in example 4, I might choose $a = 1$, $b_1 = 2$ and $b_2 = 0$, and obtain the new desirability matrix

0	1
1	0

and I could be confident that whatever value p may have in my probability matrix, the outcome of the deliberation will be the same with the new desirability matrix as it would have been with the old.

To see this, note that the original desirability matrix yields estimated desirabilities of

$$(p)(-2) + (1 - p)(1)$$

for washing, and

$$(p)(-1) + (1 - p)(0)$$

for not washing. Simplifying these expressions, we find that the estimated desirability of washing is

$$1 - 3p$$

and the estimated desirability of not washing is

$$-p .$$

These estimated desirabilities are equal when

$$1 - 3p = -p .$$

Solving for p, we find that washing and not washing are equally good acts when

$$p = .5 ,$$

in which case both acts have desirability $-.5$. If p is less than .5, the estimated desirability of washing is greater than that of not washing, while if p is greater than .5, the estimated desirability of washing is less than that of not washing.

Now when the same calculations are carried out for the new desirability matrix, the results are essentially the same. The estimated desirabilities of washing and not washing are found to be $1 - p$ and p. These are equal when $p = .5$; the estimated desirability of washing is greater than that of not washing when p is less than .5; and the situation is reversed when p is greater than .5. Then whatever value p may have, the new desirability matrix yields the same preference ranking as the old.

In no way does this result depend on the particular numbers that appeared in the original desirability matrix, or on the fact that the matrix had two rows and two columns. The general situation can be described as follows.

> Suppose that a deliberation involves a desirability matrix with m rows and n columns, and that all rows of the probability matrix are the same. Then we may choose any positive number a, and any n numbers $b_1, b_2, \ldots, b_n$; multiply all entries in the desirability matrix by a; add the number b_1 to each entry in the first column; add the number b_2 to each entry in the second column, and so on; and add the number b_n to each entry in the nth column. The result will be a new desirability matrix which yields the same preference ordering as the original.

In proving this claim we may suppose that $a = 1$, since we already know that when all entries in a desirability matrix are multiplied by the same positive number, the result is an equivalent desirability matrix. And we can reason in terms of an arbitrary *pair* of rows of the desirability matrix instead of considering all m rows; for if the transformation preserves the relative sizes of the expected desirabilities that correspond to each pair of rows, it will preserve the relative sizes of the expected desirabilities that correspond to all m rows.

Consider two rows

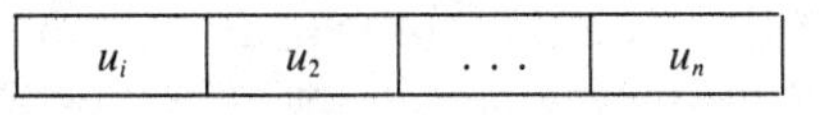

and

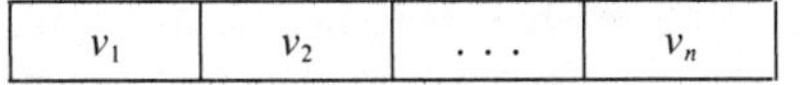

of the desirability matrix for which the corresponding rows of the probability matrix are identical:

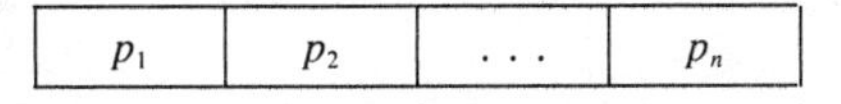

The estimated desirabilities that correspond to the two rows of the desirability matrix are u and v, where

$$u = p_1u_1 + p_2u_2 + \ldots + p_nu_n$$

$$v = p_1v_1 + p_2v_2 + \ldots + p_nv_n .$$

If we transform the desirability matrix in the indicated manner, the two rows become these

$u_1 + b_1$	$u_2 + b_2$	$\ldots$	$u_n + b_n$

$v_1 + b_1$	$v_2 + b_2$	$\ldots$	$v_n + b_n$

and the corresponding desirability estimates become u' and v' as follows:

$$u' = p_1(u_1 + b_1) + p_2(u_2 + b_2) + \ldots + p_n(u_n + b_n)$$

$$v' = p_1 (v_1 + b_1) + p_2 (v_2 + b_2) \ldots + p_n(v_n + b_n) .$$

Multiplying through and collecting terms in these expressions, we find that

$$u' = u + p_1b_1 + p_2b_2 + \ldots + p_nb_n$$

$$v' = v + p_1b_1 + p_2b_2 + \ldots + p_nb_n .$$

Then u' is obtained by adding to u a certain number b, where

$$b = p_1 b_1 + p_2 b_2 + \ldots + p_n b_n$$

and v' is obtained by adding that same number to v:

$$u' = u + b \, , \quad v' = v + b \, .$$

Therefore, it must be that u' is greater than v' if u is greater than v; that u' is less than v' if u is less than v; and that u' equals v' if u equals v. The preference ranking of the two acts cannot be disturbed by the transformation of the desirability matrix.

We can now keep the promise made in connection with example 8 of chapter 1, where we undertook to explain why the deliberation about swimming had a conclusive outcome in spite of the fact that the desirability matrix was incompletely specified as follows.

	0 days of good weather	1 day of good weather	2 days of good weather
Buy a weekend ticket	$x - 3$	$y - 3$	$z - 3$
Pay admission daily	x	$y - 2$	$z - 4$

The explanation is that since the two rows of the probability matrix are the same, we can choose any three numbers, b_1, b_2, and b_3, and add them to the entries in the three columns of the desirability matrix without affecting the outcome of the deliberation. This yields a matrix of form

$x - 3 + b_1$	$y - 3 + b_2$	$z - 3 + b_3$
$x + b_1$	$y - 2 + b_2$	$z - 4 + b_3$

Among the possible values of the three b's is the set

$$b_1 = -x \quad b_2 = -y \quad b_2 = -z \, ,$$

which yields a definite desirability matrix,

-3	-3	-3
0	-2	-4

in which no variables appear. The original incompletely specified matrix must yield a definite preference ranking of the acts because it is guaranteed to yield the same preference ranking as the completely specified matrix just obtained.

2.5 Problems

1

Which of the following desirability matrices are equivalent to each other?

0	1
2	−1

1	0
−1	2

1	2
3	0

−1	0
1	−2

2

The two desirability matrices

1	2
3	5

0	0
1	1

are equivalent in spite of the fact that neither can be obtained from the other by a transformation which changes each entry d, into a new entry of form $ad + b$. Explain.

3

Transform the desirability matrix of example 1 into an equivalent desirability matrix in which the smallest entry is 0 and the largest entry is 1.

4

In section 1.6, transform the desirability scales for the prisoners' dilemma (problem 12), Newcomb's problem (problem 14) and Fisher's problem (problem 15) so that in each case the smallest entry in the desirability matrix is 0 and the next-to-smallest is 1.

5

Which pairs of the following desirability matrices can be relied upon to yield common preference orderings whenever a common probability matrix is used in which both rows are the same?

1	0
0	1

1	0
0	2

1	1
0	2

2	−2
0	2

6 Regret

Replace each entry in a desirability matrix by the amount by which it falls short of the highest entry in its column. The result is called the *regret matrix* corresponding to the given desirability matrix. Thus, the regret matrix corresponding to the desirability matrix in example 1 is

0	2	0
1	0	2
.5	1	2

The estimated regret associated with a certain act is a weighted average of the entries in that act's row of the regret matrix, where the weights are the entries in that act's row of the probability matrix. Under what general circumstances can the principle *perform an act of minimum estimated regret* be relied upon to yield the same decisions as the Bayesian principle of deliberation?

7 Relief

Replace each entry in a desirability matrix by the amount by which it exceeds the lowest entry in its column. The result may be called the *relief matrix,* corresponding to the given desirability matrix. Thus, the relief matrix corresponding to the desirability matrix of example 1 is

1	0	2
0	2	0
.5	1	0

Under what general circumstances can the principle *perform an act of maximum estimated relief* be relied upon to yield the same decisions as the Bayesian principle of deliberation? Give an example in which the Bayesian method, the method of problem 6, and the method of this problem all lead to different decisions.

3

Ramsey's Theory

We have seen how numerical probabilities and desirabilities are used in Bayesian deliberation (chapter 1), and we have seen how one desirability scale may be replaced by another without changing the results of any deliberations (chapter 2); but only vague hints have been given about ways of discovering the numerical probabilities and desirabilities that constitute the data for deliberation. This situation will now be remedied in a preliminary way. Later, we shall take a longer look at the problem from a different point of view.

3.1 From Desirabilities to Probabilities

It is clear from problem 2 in section 1.6 that if we know what numerical desirabilities an agent attributed to the consequences in a deliberation, and we know that he liked two of the acts equally well, we may be able to discover the probabilities that he attributed to the relevant conditions.

Example 1: Using desirabilities to discover probabilities
In problem 2 of section 1.6, the consequence matrix was

	San Francisco fogged in	San Francisco not fogged in
Train	Trip takes 8 hours	Trip takes 8 hours
Plane	Trip takes 15 hours	Trip takes 3 hours

It seemed plausible to use the negatives of the trip times as desirabilities. This gave the desirability matrix

−8	−8
−15	−3

which we might have simplified by adding 15 to each entry, getting the equivalent desirability matrix

7	7
0	12

Since the agent knows that his choice of plane or train does not affect the weather in San Francisco, his probability matrix must have the form

p	$1 - p$
p	$1 - p$

in which the two rows are identical, but where p, the subjective probability that San Francisco is fogged in, is unknown. Using this probability matrix with the simplified desirability matrix, we find that the expected desirability of taking the train is 7, and the expected desirability of taking the plane is $12 - 12p$. Now if the agent finds that he is indifferent between the two acts, we know that

$$7 = 12 - 12p ,$$

so that (solving this equation) the agent must be attributing probability

$$p = 5/12$$

to the possible condition that San Francisco is fogged in.

3.2 From Probabilities to Desirabilities

It is equally clear that if we know the probabilities that an agent attributed to the relevant conditions in a deliberation, and we know that he liked two of the acts equally well, we may be able to discover what the numerical desirabilities must have been.

Example 2: Using probabilities to discover desirabilities
In problem 5 of section 1.6, the probability matrix was

	Die before the age of 65	Live to age 65 or more
Smoke 2 or more packs of cigarettes a day	.41	.59
Smoke only pipes and cigars	.25	.75

and the desirability matrix was assumed to have the form

$d + c$	$d + c + l$
d	$d + l$

where c is a measure of the agent's preference for cigarettes over pipes and cigars, and l is a measure of his preference for living to age 65 or more over dying before age 65. These increments in desirability were (perhaps rashly) assumed to be independent in the sense that the gain in desirability due to smoking cigarettes instead of only pipes and cigars was taken to have the same value c, in the longer life as in the shorter, and the gain in desirability due to living longer was taken to have the same value l, whether the agent smokes cigarettes, or only pipes and cigars. The desirability matrix can be simplified by subtracting d from each entry:

c	$c + l$
0	l

The estimated desirabilities of the two acts are then as follows:

$$c + .59l$$

$$.75l$$

If the agent finds that he is indifferent between the two acts, we know that these expressions are equal and, solving the resulting equation, we find that

$$c = .16l \, .$$

Substituting $.16l$ for c in the simplified desirability matrix we have

$.16l$	$1.16l$
0	l

Since the agent would rather live longer than not. l is positive, and accordingly, we can divide each entry by l (multiply each entry by $1/l$) to get an equivalent desirability matrix,

.16	1.16
0	1

in which there are no unknowns. We have now located the four consequences on a desirability scale as shown in table 3.1.

Table 3.1

1.16	Live to 65 or more, smoking 2 or more packs of cigarettes a day.
1.00	Live to 65 or more, smoking only pipes and cigars.
0.16	Die before 65, smoking 2 or more packs of cigarettes a day.
0.00	Die before 65 in spite of smoking only pipes and cigars.

3.3 The von Neumann–Morgenstern Method

There was a large element of luck in our discovery of the agent's numerical desirabilities in example 2: the agent had to find himself indifferent between the two acts. In anything that deserves to be called a method for getting desirabilities out of probabilities, this element of luck must be eliminated. The general problem of measuring desirabilities has this form. We are presented with a preference ordering in which a consequence C is ranked between two given consequences, A and B.

$$B$$

$$C$$

$$A$$

We seek to find exactly where C lies in the desirability interval from A to B. We can do this if a deliberation can be found in which the consequence matrix is as follows; in which the probabilities of the two conditions are known and are independent of which act is performed; and in which the agent is indifferent between the two acts.

	Condition 1	Condition 2
Act 1	B	A
Act 2	C	C

The probability matrix then has form

p	$1 - p$
p	$1 - p$

;

the desirability matrix may be taken to have the form

1	0
x	x

where the desirabilities of the extreme consequences A and B are arbitrarily taken to be 0 and 1; the acts have estimated desirabilities p and x; and since the acts are liked equally well, p and x are equal. The desirability x of the consequence C can thus be determined as equal to the probability p that the agent attributes to condition 1. A deliberation of the required sort can always be concocted by devising a suitable gamble to play the role of act 1, as in the following example. This method of measurement was discovered by F. P. Ramsey and rediscovered by von Neumann and Morgenstern, through whose work it came to play its current role in economics and statistics.

Example 3

The agent ranks the prospects of having tuna, ham, and egg sandwiches for lunch today as follows:

tuna

ham

egg

Suppose that, in fact, ham is 1/4 of the way from egg to tuna on his desirability scale. This will be shown by the fact that the agent is indifferent between the two acts in the following deliberation, in which the conditions refer to the result of blindly drawing a card from a standard deck.

	A heart is drawn	A heart is not drawn
Have tuna if a heart is drawn, egg if not	tuna	egg
Have ham	ham	ham

Had ham been located elsewhere in the interval from egg to tuna, this would have been shown by the agent's indifference between the two acts in a similar deliberation.

In general, we can calibrate the desirability scale between two consequences as finely as need be, by using appropriate gambles. The desirability interval from egg to tuna can be divided into 52 equal segments by gambles relating to an ordinary pack of playing cards, and the desirabilities of all consequences that lie between egg and tuna would thereby be determined within intervals of length 1/52. Thus, there are 13 hearts and 12 face cards, so that if having a ham sandwich had been found to lie between the gambles tuna if a heart is drawn, egg if not

and

tuna if a face card is drawn, egg if not

in the preference ranking, the desirability of having a ham sandwich would thereby have been located within the thirteenth of the 52 equal intervals from egg to tuna in the desirability scale, being greater than 12/52 but less than 13/52.

3.4 Ethical Neutrality; Probability 1/2

To calibrate an agent's desirability scale by means of gambles in this way, we need a method for discovering the probabilities that he attributes to at least some of the conditions he might gamble on: the von Neumann–Morgenstern technique allows us to measure desirabilities in terms of preferences and probabilities, not in terms of preferences alone. Such a method does exist; it was discovered by F. P. Ramsey in 1926 and used by him as part of a general solution to the problem of measuring subjective probabilities and desirabilities. (But see section 3.9: in the present account, a certain gratuitously awkward feature of Ramsey's theory is replaced by the corresponding feature of a theory worked out by Donald Davidson and Patrick Suppes ca. 1954.)

Ramsey observed that there is a simple test for determining whether what he called an *ethically neutral* condition has probability 1/2. A condition is ethically neutral in relation to a particular agent and a particular consequence if the agent is indifferent between having that consequence when the condition holds and when it fails. We shall speak of unqualified ethical neutrality when the condition is ethically neutral for the agent relative to all of the consequences that are under discussion. Thus, in example 3, the condition that a heart is drawn is ethically neutral since the agent's appetite for the different kinds of sandwiches would not be affected by the outcome of the drawing. However, if the agent is loath to eat meat on Fridays, and does not remember what day it is, the condition that today is Friday might not be ethically neutral for him. If he were convinced that today is not Friday, his preference ranking of consequences might be

tuna

ham, cheese

egg

where he is indifferent between ham and cheese; but if he were convinced that today is Friday, ham would presumably move to a new bottom rank:

tuna

cheese

egg

ham.

Since he is in doubt about what day it is, his actual preference ranking may differ from either of these two; perhaps it would be

tuna

cheese

ham

egg

if he strongly doubts that today is Friday and strongly prefers ham over egg.

We can now state Ramsey's criterion for recognizing when an ethically neutral condition has probability 1/2.

> Suppose that *A* and *B* are consequences between which the agent is not indifferent, and that *N* is an ethically neutral condition. Then *N* has probability 1/2 if and only if the agent is indifferent between the following two gambles.
>
> *B* if *N*, *A* if not
>
> *A* if *N*, *B* if not

Thus, if *A* is egg, if *B* is tuna, and if *N* is the condition that a red card is drawn, the two gambles are these:

tuna if a red card is drawn, egg if not

egg if a red card is drawn, tuna if not.

Imagine that you do not know how probable the agent thinks it is that a red card will be drawn, but that you do know that he likes the two gambles equally well. The desirability of tuna is z; the desirability of egg is x; and the probability that a red card will be drawn is an unknown number, p. The estimated desirabilities of the respective gambles are

$$pz + (1 - p)x$$
$$px + (1 - p)z$$

and you know that those are equal:

$$pz + (1 - p)x = px + (1 - p)z .$$

Simplifying, the equation becomes

$$2pz - 2px = z - x ,$$

and dividing both sides by $2(z - x)$, which cannot be 0, since tuna is preferred to egg, we find that indeed

$$p = 1/2 .$$

Then from the facts (1) that consequences *A* and *B* do not have the same preference ranking, (2) that *N* is ethically neutral, and (3) that the gambles

B if *N*, *A* if not

and

A if *N*, *B* if not

do have the same preference ranking, it does follow that *N* has probability 1/2. For a better understanding of why this is so, consider what happens when the ethically neutral condition *N* has a probability other than 1/2. In particular, consider the conditions

an ace is drawn,

a face card is drawn,

a 5, 6, 7, 8, 9, or 10 is drawn

which have probabilities 1/13, 3/13, and 6/13. The expected desirabilities of the three pairs of gambles are then symmetrically located on opposite sides of the midpoint of the desirability interval from *A* to *B*, e.g., from egg to tuna. This is illustrated in table 3.2 where the desirability (x) of egg is taken to be 0 and the desirability (z) of tuna is taken to be 1. As

Table 3.2

1	tuna
12/13	egg if an ace is drawn, tuna if not
11/13	
10/13	egg if a face card is drawn, tuna if not
9/13	
8/13	
7/13	egg if a 5, 6, 7, 8, 9, or 10 is drawn, tuna if not
6/13	tuna if a 5, 6, 7, 8, 9, or 10 is drawn, egg if not
5/13	
4/13	
3/13	tuna if a face card is drawn, egg if not
2/13	
1/13	tuna if an ace is drawn, egg if not
0	egg

the probability of the condition N approaches 1/2, the estimated desirabilities of the gambles

tuna if N, egg if not

egg if N, tuna if not

approach each other from opposite sides of the midpoint.

3.5 Calibrating the Desirability Scale

Now Ramsey calibrates the desirability scale between consequences A and B, where B is preferred to A, as follows. He finds an ethically neutral condition N which has probability 1/2 and identifies the midpoint of the scale as the estimated desirability of the gamble

A if N, B if not.

To identify the 1/4 and 3/4 points on the scale, he finds a prospect (say, C) which is liked exactly as well as that gamble and identifies the 1/4 point on the scale as the estimated desirability of a second gamble

A if N, C if not;

and he identifies the 3/4 point as the estimated desirability of a third gamble

C if N, B if not.

Each of these quarters of the scale can then be split by the same procedure, as can each of the resulting eighths, etc., to obtain as fine a calibration as necessary. The method is illustrated in table 3.3, where N is the condition that a red card is drawn; where A (=egg) is assigned desirability 0; B (= tuna) is assigned desirability 1; C (=ham) is supposed to be liked exactly as well as a gamble on N between egg and tuna; liverwurst is found to be liked exactly as well as the gamble that marks the 1/4 point; and where a gamble on N between liverwurst and ham can therefore be used to mark the 3/8 point.

Table 3.3

1	tuna
7/8	
3/4	ham if a red card is drawn, tuna if not
5/8	
1/2	ham; egg if a red card is drawn, tuna if not
3/8	liverwurst if a red card is drawn, ham if not
1/4	liverwurst; egg if a red card is drawn, ham if not
1/8	
0	egg

3.6 Measuring Probabilities

In this way, using gambles on a single ethically neutral condition of probability 1.2, Ramsey is able to calibrate the desirability scale between any two consequences. Now if on such a scale the gamble

$$A \text{ if } M, B \text{ if not}$$

has desirability g, and the consequences

$$A \text{ when } M \text{ is true,} \quad B \text{ when } M \text{ is false}$$

have desirabilities e and f, respectively, then we can express the desirability of the gamble in terms of those of the two consequences as follows

$$g = pe + (1 - p)f ,$$

where p is the subjective probability of M. (Observe that M need not be ethically neutral, and so e and f need not be the desirabilities of A and B themselves.) Multiplying through by f and collecting terms, the equation becomes

$$p(e - f) = g - f .$$

If the agent is not indifferent between A when M is true and B when M is false, i.e., if e and f are distinct, we then have

$$p = \frac{g - f}{e - f} .$$

as the probability that the agent attributes to M.

Example 4

In table 3.3, let A be tuna and let B be egg. Suppose that the sandwiches are all made with white bread or are all made with rye, but you don't know which. Let M be the proposition that rye is used. Suppose that tuna on rye is liked exactly as well as ham, so that $e = 1/2$; that egg on white bread is liked exactly as well as the gamble

liverwurst if a red card is drawn, egg if not,

so that $f = 1/8$; and that the gamble

tuna if rye is used, egg if not

has desirability $g = 1/4$, being liked exactly as well as liverwurst. It follows that the subjective probability of the proposition that rye is used is

$$p = \frac{(1/4) - (1/8)}{(1/2) - (1/8)} = \frac{1}{3} .$$

3.7 Conclusion

Our original problem was that too great a variety of desirability assignments were compatible with the given preference ranking of *consequences:* there was no assurance that two such desirability assignments would be equivalent in the sense of always leading, via a common probability matrix, to one and the same preference ranking of *acts*. Ramsey solved the problem by extending the preference ranking to include gambles between the given consequences.

Given the extended preference ranking, the subjective probabilities of all conditions on which the agent might gamble are determined by Ramsey's method, and the desirabilities of all consequences are determined as definitely as necessary for purposes of Bayesian deliberation; for *if two desirability assignments are both compatible with the extended preference ranking, they must be equivalent.* This equivalence follows from the fact that according to Ramsey's method, once desirabilities have been assigned to some pair of differently ranked consequences, the desirabilities of all other consequences are determined. Thus, once desirabilities 0 and 1 were assigned to egg and tuna in our example, the other desirabilities were uniquely determined by the preference ordering of consequences and gambles between consequences.

Of course, for Ramsey's method to work, the extended preference ranking of consequences and gambles between consequences must satisfy certain conditions of coherence: the ranking must be such as might have been obtained in the Bayesian manner from a previously given assignment of probabilities to conditions and of desirabilities to consequences.

Example 5

Suppose that this is a portion of someone's extended preference ranking:

(1) tuna if rye is used, egg if not

(2) tuna when rye is used

(3) egg when rye is not used

Then the ranking is incoherent in the sense that no desirability assignment is compatible with it. For if the desirabilities of the prospects (1), (2), and (3) are taken to be g, e, and f, must have

(4) $$g > e > f$$

for compatibility with the given portion of the preference ranking; and if the probability of rye being used is p, the Bayesian principle specifies that the estimated desirability of prospect (1) is

$$g = pe + (1 - p)f \, . \tag{5}$$

But since p is a number in the interval from 0 to 1, the value of the right-hand side of (5) must lie somewhere in the interval from e to f: if p is 0 the value is f, if p is 1 the value is e, and if p is between 0 and 1 the value is between e and f. Then conditions (4) and (5) contradict each other, and the preference ranking is incoherent.

Furthermore, if Ramsey's method is to allow arbitrarily precise measurement of desirabilities, the preference ranking must have certain special characteristics. Thus, if we are to be able to calibrate the desirability scale with the aid of just one ethically neutral condition N of probability 1/2, it will be necessary for there to be infinitely many consequences; in particular, for every gamble on N there must be a consequence which is liked exactly as well. Thus, in table 3.3, in order to mark the 3/8 point of the scale, we needed consequences (ham, liverwurst) which were liked exactly as well as the gambles that marked the 1/2 and 1/4 points. The 3.8 point was then marked by the gamble

liverwurst if a red card is drawn, ham if not

between those consequences.

This exposition of Ramsey's work has been informal and fragmentary. Examples have been given of the conditions that a preference ranking must satisfy in order to determine a unique probability assignment and to determine a desirability assignment which is unique once desirabilities have arbitrarily been assigned to two consequences of different ranks. But no attempt has been made to give a complete list of such conditions, nor have we attempted to deal with preference rankings in full generality. In particular, the method sketched here will yield a desirability assignment if there is a best-liked consequence and a least-liked consequence, to which it may be convenient to suppose that desirabilities 1 and 0 are assigned, as in table 3.3. But the method must be slightly modified to deal with cases in which, no matter how high or low a consequence may be in the preference ranking, there is always a better-liked consequence and a worse-liked consequence. This may be the case even though all desirabilities lie within some finite interval, e.g., between -1 and 1.

Example 6

Suppose that situations in which I am to be paid 0, 1, 2, 3, . . . , n, . . . dollars have desirabilities

$$0, 1/2, 3/4, 7/8, \ldots, (2^n - 1)/2^n, \ldots,$$

where each number halves the distance from its predecessor to 1. Also, suppose that the desirability of the situation in which I am to pay n dollars is just the negative of the desirability of the situation in which I am to be paid n dollars. If n is allowed to take on all finite positive values, no consequence is at the top or at the bottom of this preference ranking, and yet the desirabilities of all prospects are less than 1 and greater than -1.

3.8 Problems

1

The agent's preference ranking of consequences and gambles is shown below, where *M* and *N* are ethically neutral propositions, and where *A*, *B*, *C*, and *D* are consequences. Assume that the desirability of *A* is 0 and that the desirability of *C* is 1. Find the desirabilities of *B* and of *D*, and find the probabilities of *M* and of *N*.

D

C; *D* if *N*, *A* if not

B; *C* if *N*, *A* if not; *A* if *N*, *C* if not; *D* if *M*, *A* if not

A

2

Find the probability of *M*, given that *M* and *N* are ethically neutral propositions and that the preference ranking is as follows:

A

B; *A* if *N*, *C* if not; *C* if *N*, *A* if not

C; *B* if *N*, *E* if not; *B* if *M*, *D* if not

D; *C* if *N*, *E* if not

E

3 The Absent-Minded Catholic

The agent is loath to eat meat on Fridays; is sure that today is Thursday or Friday, but doesn't know which; and is uncertain about the percentage of face cards in a deck. It is evident from an examination of his entire preference ranking of consequences that the condition *a face card is drawn* is ethically neutral for him. A portion of his preference ranking of consequences and gambles between consequences is as follows.

Tuna on Thursday. Tuna on Friday. Tuna today.
Ham on Thursday. Egg today if a face card is drawn, tuna today if not.
Tuna today if a face card is drawn, egg today if not.

Egg on Thursday. Egg on Friday. Egg today. Ham on Thursday if a face card is drawn, ham today if not.
Ham today. Ham on Friday if a face card is drawn, egg today if not.
Ham on Friday.

Assign desirability 0 to *egg today,* desirability 1 to *tuna today.* Discover the desirabilities of the remaining consequences and gambles listed above, and discover the probabilities that the agent attributes to the two conditions, *today is Friday* and *a face card is drawn.* Hint: *ham today* can be construed as the gamble

Ham on Friday if today is Friday, ham on Thursday if not.

4

What can be concluded about the probability of the ethically neutral proposition N, when the preference ranking is as follows?

A; B if N, A if not.

B; A if N, B if not.

5

What are the desirabilities of A and of B when the preference ranking is as follows and N is ethically neutral?

A; A if N, B if not; B if N, A if not.

What can be concluded about the probability of N?

6

Assume that the desirabilities of receiving various amounts of money are as shown in figure 3.1, and that the agent has the usual notions about the probabilities of drawing hearts and red cards (from a standard deck). What is the preference ranking of the following consequences and gambles?

(a) Receive $1
(b) Receive $2
(c) Receive $3
(d) Receive $0 if a heart is drawn, $4 if not.
(e) Receive $4 if a heart is drawn, $0 if not.
(f) Receive $0 if a red card is drawn, $4 if not.

7

The situation shown in figure 3.1 is one in which the agent sorely needs a dollar or so and has a less pressing need for larger amounts. In the situation shown in figure 3.2, the agent sorely needs about $3. What is the preference ranking of (a–f) in problem 6 when the situation is as described in figure 3.2?

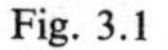

Fig. 3.1 Fig. 3.2

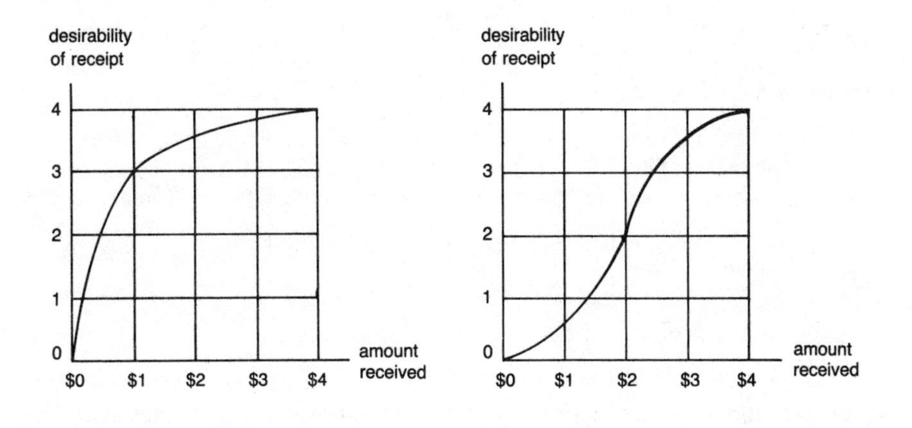

3.9 Notes and References

Ramsey's theory is set forth in an essay, "Truth and Probability," that was published only posthumously, in a collection of his writings edited by R. B. Braithwaite: *The Foundations of Mathematics and other Logical Essays* (London: Routledge and Kegan Paul; New York: Harcourt Brace, 1931). It also appears in a somewhat modified version of the 1931 collection, edited by D. H. Mellor, *Foundations* (London: Routledge and Kegan Paul, 1978) and in in a collection of essays by various authors, *Studies in Subjective Probability,* 2d ed., edited by Henry Kyburg, Jr., and Howard Smokler (Huntington, New York: Robert E. Krieger, 1980).

Ramsey envisages a bizarre caricature of a psychological test for identifying certain propositions to which a subject attributes probability 1/2. It requires a belief on the subject's part that the experimenter has the power of the Almighty and so can effectively offer as options, and as possible outcomes of gambles, complete possible courses of the world α, β, . . . , i.e., the "different possible totalities of events between which our subject chooses—the ultimate organic unities" (pp. 78–79). Following Ludwig Wittgenstein, *Tractatus Logico-Philosophicus* (London: Routledge and Kegan Paul; New York: Harcourt Brace, 1922, which he translated), Ramsey postulated the existence of *atomic* propositions, each of which can be true or false quite independently of any or all of the others, and which have the characteristic that any proposition whatever is some truth functional compound of atomic ones. Now Ramsey defines ethical neutrality as follows:

> An atomic proposition *p* is called ethically neutral if two possible worlds differing only in regard to the truth of *p* are always of equal value; and a non-atomic proposition *p* is called ethically neutral if

> all its atomic truth-arguments are ethically neutral. [*Foundations*, p. 79]

He remarks in a footnote, "I assume here Wittgenstein's theory of propositions; it would probably be possible to give an equivalent definition in terms of any other theory.". He continues:

> We begin by defining belief of degree 1/2 in an ethically neutral proposition. The subject is said to have belief of degree 1/2 in such a proposition p if he has no preference between the options
>
> (1) α if p is true, β if p is false, and
>
> (2) α if p is false, β if p is true ,
>
> but has a preference between α and β simply.

But of course, p will have a definite truth value in (complete) possible world α, and in β, too. Ramsey remarks in a footnote to this passage, "α and β must be supposed so far undefined as to be compatible with both p and not-p." The point is that if (1) and (2) are to be gambles, the prizes α and β are to be awarded depending on whether p is true or false in the real world—the course of which is so far unknown to the subject at the time he expresses his preference between (1) and (2) as to leave him uncertain of their outcomes.

The thing makes an abstract sort of sense if you think of the business of specifying a possible world as simply a matter of specifying, for each atomic proposition, whether it is to be true or false in the world in question. Each such specification can be made in either way, independently of how the truth values of the other atomic propositions have been or will be specified. Then if α is any possible world, and p is any atomic proposition, there will be a possible world—call it "$\alpha + p$"— that is just like α except (perhaps) that p is true in it. In other words, $\alpha + p$ is simply α if p is true in α, while if p is false in α, $\alpha + p$ is the world in which p is true but all other atomic propositions have the same truth values they have in α. Similarly, $\alpha - p$ is α itself in case p is false in α, while if p is true in α, $\alpha - p$ is to be the world in which p is false but in which all other atomic propositions have the same truth values as in α. Using this notation, Ramsey's options (1) and (2) can be represented as follows:

(1*) $(\alpha + p)$ if p is true, $(\beta - p)$ if p is false.

(2*) $(\alpha - p)$ if p is false, $(\beta + p)$ if p is true.

Ramsey remarks, "This comes roughly to defining belief of degree 1/2 as such a degree of belief as leads to indifference between betting one way and betting the other for the same stakes," and explains his use of possible worlds as stakes, "This is, of course, a very schematic version of the situation in real life, but it is, I think, easier to consider it in this form" (*Foundations*, pp. 79, 80). Few would share that opinion.

Following Donald Davidson and Patrick Suppes, "A Finitistic Axiomatization of Subjective Probability and Utility," *Econometrica* 24 (1956): 264–75, and *Decision Making* (1957; Midway reprint, Chicago: University of Chicago Press, 1977), chapter 1, the treatment of Ramsey's theory in this chapter simply sets the business about possible worlds and atomic propositions aside as a useless complication. But where Davidson and Suppes leave open the character of their stakes, I take them to be the same sorts of things to which probabilities are attributed, i.e., truths of statements—propositions, one might as well say, as long as it is understood that we are not committed to Wittgenstein's account of them. In place of Ramsey's worlds α, β, I have propositions q, r—propositions that, like p, need not be atomic. (I do not suppose there to be any such things as atomic propositions.) There is now no need to replace Ramsey's (1) and (2) by (1*) and (2*) as above.

Example 6

Consider the option at the 12/13 mark in table 3.2:

Egg if an ace is drawn, tuna if not .

I take this to be an option of Ramsey's form (1), except that in place of his worlds α, β, I have propositions q, r:

$$q \text{ if } p,\ r \text{ if not.}$$

Here, p, q, and r are the propositions that

p = The card that will now be drawn is an ace.

q = I shall have an egg sandwich for lunch.

r = I shall have a tuna sandwich for lunch.

For more about this, see section 10.3.

The von Neumann–Morgenstern method appeared in *Theory of Games and Economic Behavior* by John von Neumann and Oskar Morgenstern (Princeton, N.J.: Princeton University Press, 1944), pp. 15–31; 2d ed. (1947), pp. 617–32. For a brief, simple introduction, see Herman Chernoff and Lincoln E. Moses, *Elementary Decision Theory* (New York: Wiley), pp. 80–83, 350–52.

In the first five chapters of *Foundations of Statistics* (1954), Savage sets forth a theory with affinities to Ramsey's, through which (over the next decade or so) subjective probability theory gained a certain respectability among statisticians. Savage's exposition is highly technical, and difficult. For a brief, clear exposition of the central ideas, see R. Duncan Luce and Howard Raiffa, *Games and Decisions* (New York: Wiley, 1957), pp. 302–4.

4

Propositional Attitudes

Ramsey attributes desirabilities to consequences but attributes probabilities to entities of another sort: to possible conditions or (as we shall also say) propositions. The needed links between the two domains are provided by gambles, i.e., hybrid entities whose positions in the preference ranking correspond to their estimated desirabilities and thereby show what probabilities the agent attributes to the conditions gambled upon, and what desirabilities he attributes to the consequences (the possible outcomes of the gambles).

In this and the following six chapters, we shall develop an alternative to Ramsey's system: a theory of preference which is *unified* in the sense that it attributes probabilities and desirabilities to the same objects, and *noncausal* in a sense that will be explained as the theory is developed. This chapter and the next are devoted to philosophical preliminaries and to an exposition of such elementary facts about logic, probability, and desirability as are needed for the theory of preference.

4.1 Belief and Desire

To believe that it will rain tomorrow is to have a particular attitude toward the proposition that it will rain tomorrow; to desire that it rain tomorrow is to have another sort of attitude toward the same proposition. When the form of ordinary talk suggests that the objects of desire are things other than propositions, it is always possible, and often useful, to suppose that various corresponding propositions are the actual objects of the attitude. To desire peace, or a certain job, or the love of a good woman, or a ham sandwich, is to desire that one proposition or another hold (= obtain = be the case = be true): that there be peace, or that the desirer get the job, or some good woman's love, or a ham sandwich now.

Of course we do speak of jobs, sandwiches, women, and men as objects of desire, and these things are not propositions. But in such cases the highly flexible notion of having something is linked with the notion of desiring it: to desire x is to desire that one have x (in the appropriate sense of "have") so that what is desired is, after all, a proposition. The object here is not to reform ordinary talk, but to interpret it. I do not propose to substitute clumsy, explicitly propositional locutions for common idioms, but I do propose to develop a theory of preference in which desires are uniformly taken to be for the truth of propositions. If it is correct to interpret ordinary talk about desire in this way, there is no real restriction in supposing that it is only propositions that are ranked by the preference ordering.

4.2 Justifying the Special Addition Law

Subjective probability is partial belief. To say that an agent attributes (subjective) probability .7 to the proposition that it will rain tomorrow is to say that his degree of belief in that proposition is .7; and this in turn means (roughly) that he would be just willing to pay $.70 in order to receive $1 if it rains tomorrow and nothing if it doesn't. This is a rough characterization: I take it that the meaning of the probability statement is given more accurately by Ramsey's theory, and by the theory that will be developed in the following chapters, but for all its roughness, the characterization is a useful one. One of its most striking uses is to justify the assumptions of the elementary theory of probabilities. The argument is as follows.

Grant that to attribute probability p to a proposition is to imply that it would be fair to pay p dollars to get one dollar if the proposition is true, and nothing if it is false. Then the following assumption, which we shall call the *special addition law*, can be justified.

> If a set of conditions are mutually exclusive and collectively exhaust the possibilities, their probabilities must add up to 1.

For definiteness, consider the case where there are just three conditions; the argument would be entirely parallel for a set of two conditions, or four, or a thousand. In particular, consider the following set of three conditions:

R = it will rain but not snow tomorrow;

S = it will snow but not rain tomorrow; and

N = it will neither rain nor snow tomorrow.

Suppose the agent attributes probabilities r, s, n to these three propositions. Then he thinks that each of the following arrangements is fair.

(i) Pay r dollars to get \$1 if R, nothing if not.

(ii) Pay s dollars to get \$1 if S, nothing if not.

(iii) Pay n dollars to get \$1 if N, nothing if not.

Since he thinks that each arrangement by itself is fair, he thinks it fair to make all three arrangements together. Suppose he does so. Then see what the results of the three arrangements will be in each of the three (mutually exclusive) cases that can arise. This is done in table 4.1, where the cases are listed at the left, the gains from the various arrangements in the various cases are listed in the middle, and the net gains from all three arrangements in each case are listed at the right. Losses are indicated as negative gains.

Table 4.1

Case	Gain from (i)	Gain from (ii)	Gain from (iii)	Net Gain
R	$1 - r$	$-s$	$-n$	$1 - r - s - n$
S	$-r$	$1 - s$	$-n$	$1 - r - s - n$
N	$-r$	$-s$	$1 - n$	$1 - r - s - n$

Now the net gain from all three arrangements is the same, no matter which of the three possible cases actually holds. Therefore the payment could have been made before finding out what the weather will be tomorrow: to make all three arrangements is to be certain of realizing a net gain of $1 - r - s - n$ dollars. If this amount is positive, the arrangements could not all have been fair, because the agent has the unfair advantage of winning come what may. If the amount $1 - r - s - n$ is negative, the arrangements could not all have been fair, because the agent has the unfair disadvantage of losing come what may. It is only when $1 - r - s - n = 0$ that the arrangements could all have been fair. Therefore we must have

$$r + s + n = 1$$

for fairness. In other words, the addition law must hold if the probability assignment is to be fair.

4.3 Remarks on Fairness

A probability assignment can be unfair even though the addition law holds. If the propositions R, S, and N refer to the weather in Los Angeles in July, and if r, s, and n are all 1/3 in the foregoing bets, the addition law is satisfied, but the numbers are absurdly at variance with experience. Thus, suppose the agent takes the probabilities of R, S, and N to be .01, .005 and .985. He will regard each of the arrangements (i), (ii), and (iii) as unfair if in them we have $r = s = n = 1/3$, because he takes the expected gains from the three arrangements to be

$$\text{(i)} \quad .01(1 - r) + .005(-r) + .985(-r) = .01 - r = -32¢\,,$$

$$\text{(ii)} \quad .01(-s) + .005(1 - s) + .985(-s) = .005 - s = -33¢\,,$$

$$\text{(iii)} \quad .01(-n) + .005(-n) + .985(1 - n) = .985 - n = +65¢\,.$$

This means that he takes the first two arrangements to be unfairly disadvantageous, and the third to be unfairly advantageous. In general, the agent regards an arrangement as fair if and only if the expected gain from it, computed in accordance with the probabilities that *he* attributes to the possible cases, is zero.

In the case we have been considering, the agent regards the three arrangements together as fair, since the expected net gain is the sum

$$(-32) + (-33) + (65) = 0$$

of the expected gains from the individual arrangements. However, he also must regard it as pointless to make all three arrangements, because not only is the expected net gain zero, but the actual net gain is inevitably zero, no matter which of the three possible cases R, S, N is actual. From the three arrangements together there is no chance of gain, and no chance of loss, if r, s, n add up to 1.

It is worth noting the assumptions that were used in justifying the special addition law. We had to suppose the following: (1) that if the agent takes the probability of a proposition to be p, then he thinks it fair to pay p dollars in order to receive one dollar if the proposition is true, and nothing if it is false; (2) that if the agent thinks that each of several such arrangements is individually fair, he must think it fair to make all of the arrangements together; and (3) that an arrangement or a set of simultaneous arrangements is unfair (= unfairly advantageous) if the gain or the net gain is positive in each case, and that an arrangement or a set of arrangements is unfair (= unfairly disadvantageous) if the gain or the net gain is negative in each case.

Subjective probability is partial belief; beliefs are manifest in action and in the agent's attitudes toward alternative courses of action; and in particular, subjective probabilities may be manifest in an agent's judgments about the fairness of certain risky monetary arrangements, as in supposition (1) above.

4.4 Desirability

Socrates argues (*Symposium,* 200) that Love is not a god because to desire something is to be in want of it: you cannot desire what you already have. The point is also made in contemporary dictionaries. Thus, *The Concise Oxford Dictionary* defines a desire as an *unsatisfied* appetite, longing, wish, or craving. This account of the matter seems roughly right, although it needs some refinement. Thus, to vary the example of the *Symposium,* since people do not always know when they are loved, it is entirely possible to desire someone's love when you already have it. Therefore, it seems better to say that you cannot desire what you *think* you have. Taking propositions as the objects of desire, the doctrine becomes: one who believes that a proposition is true cannot desire that it be true.

Now in the technical sense, subjective estimates of desirability are to be degrees of desire, just as subjective probabilities are to be degrees of belief. This makes for conflict with ordinary usage, for I am ordinarily said to think desirable not only things I desire but also various things that I have and know I have. It may indeed be that I desire something, and later, when I have it, find it is not as good as I had thought. This is a case where my judgments of desirability have changed as a result of experience. Perhaps the later judgment is the sounder, being based on more experience. Or perhaps not; perhaps I am in such a sorry state that I regularly despise what I have because I have it. But as we ordinarily use the term, desirability is not logically incompatible with known possession. In contrast, as we shall see in chapter 5, full belief washes out numerical (estimated, subjective) desirability, for any two propositions of probability 1 must have equal desirability.

Can I be mistaken in thinking I desire something? (The question is not whether I can misestimate something's desirability, but whether I can now be mistaken about what my current estimate is, of something's desirability.) I suppose so, and I suppose that I may discover my mistake after I have acquired the thing and come to despise it. But this way of discovering the mistake would be accidental: the experience of having the thing might remind me of attitudes I had about it before I possessed it—attitudes which, had I been more thoughtful or sensitive, I could have recognized before the acquisition. Ordinarily, when I first believe I desire something and then, acquiring it, believe I despise it, both beliefs are right. At the time I thought I desired it I did desire it, and later I came to despise it and to believe I despised it. Subjective probabilities and desirabilities here relate to a particular person at a particular time. On the whole we know what our attitudes are, but often enough, friendly or clinical observers may read our hearts better than we.

I have been at pains to point out some of the connections between the basic terms of the theory of preference and certain related terms that appear in ordinary talk. It should not be concluded from this that the theory of preference between propositions is put forth as a branch of

English linguistics, or as a branch of the common linguistics of English and classical Greek. Ordinary talk of today and of the fifth century B.C. is relevant because the ordinary practical concerns today and then are relevant, and those concerns are reflected in talk. The notions of subjective probability, subjective desirability, and preference between propositions apply to those concerns, as do the notions of desire and belief. Therefore the two sets of notions are related to each other, although in complex ways that defy summary description. It is profitable for us to begin our understanding of the new theory by describing its concepts in terms that are already familiar to us, just as it is profitable for an adult to begin the study of a foreign language by relating its terms to those of the mother tongue.

4.5 Sentences and Propositions

It does not weaken this disclaimer to point out that propositions either are linguistic entities, or at the very least, have strong affinities to certain linguistic entities. We commonly name propositions by putting the word "that" or the phrase "the proposition that" before the corresponding declarative sentences. The relation between a sentence and the proposition it expresses is thus akin to the relation between direct and indirect quotation, and this is no superficial similarity. Sentences that have the same meaning express the same proposition; thus, since the sentence

Magellan circumnavigated the globe

has the same meaning as the sentence

Magellan sailed all the way around the earth,

the proposition that Magellan circumnavigated the globe is identical with the proposition that Magellan sailed all the way around the earth. Now suppose I hear someone utter the first of these two sentences about Magellan. I can therefore say truly,

He said, "Magellan circumnavigated the globe,"

but not

He said, "Magellan sailed all the way around the earth."

Direct quotation discriminates between sentences that mean the same, but this is not so for indirect quotation. I can say,

He said that Magellan sailed all the way around the earth

as truly as I can say

He said that Magellan circumnavigated the globe.

In direct quotation, the direct object of "said" is the name of a sentence, formed by putting that sentence between quotation marks. In indirect quotation, the direct object of "said" is the name of a proposition, formed by putting the word "that" in front of a sentence that expresses it. We shall uniformly interpret "desired" and "believed" as analogous to "said" in indirect quotation. In the case of "believed," this is no mere analogy. It is good English, and true, to say

Columbus believed that the earth is round.

In the case of "desired" the analogy is not strict. It is not good English to say

Columbus desired that the earth is round,

for Columbus cannot properly be said to have desired the truth of a proposition that he believed true. The change from desire to desirability partially restores the point. It is still not good English (but it would be true if it were) to say,

For Columbus, it was desirable that the earth is round.

To meet the demands of grammar and truth, we must resort to some such clumsy locution as

From Columbus's point of view it was desirable that
it should be the case that the earth is round.

4.6 Notation

Our concern, however, is not basically with these terms of English, but with the analogous terms of the theory of preference. For brevity and clarity we shall use the notation

prob A

for

the probability that the agent attributes to the proposition *A*,

and we shall use

des A

for

the desirability that the agent attributes to the proposition *A*.

Then the symbols *prob A* and *des A* stand for numbers. We shall use Italic capitals to stand for propositions; and we shall use special symbols that correspond (roughly) to the words

not, and, or,

to form *denials, conjunctions,* and *disjunctions* of propositions. In particular, suppose that propositions A and B are expressed by English sentences, for example, by the sentences

It is raining

and

It is snowing.

Then $\bar{A}$, the denial of A, is expressed by the English sentence

It is *not* raining,

or, more cumbersomely,

It is not the case that it is raining;

AB, the conjunction of A and B, is expressed by the English sentence

It is raining *and* it is snowing;

and $(A \vee B)$, the disjunction of A and B, is expressed by the English sentence

It is raining *or* it is snowing,

or, more cumbersomely and explicitly,

It is raining *and/or* it is snowing.

Now let us put the matter still more explicitly. Suppose that we have two or more propositions; for definiteness, suppose they are the three propositions

$A, B, C,$

which are expressed by the English sentences

It is raining,
It is snowing,
It is blowing.

Their conjunction is the proposition

ABC

which asserts that all three component propositions are true: the conjunction is true if all three conjuncts, $A, B, C,$ are true, and the conjunction is false if even one of the three conjuncts is false. The disjunction of the three propositions is the proposition (note the parentheses!)

$$(A \vee B \vee C)$$

which is true if any one or more of the three disjuncts A, B, C, are true, and is false only if all three disjuncts are false. Finally, the denial of a proposition A is the proposition

$$\overline{A}$$

which is true if A is false and false if A is true.

The parentheses that enclose disjunctions serve to indicate grouping. The expression

$$A(B \vee C)$$

can only be read as the conjunction of A with $(B \vee C)$: it denotes the proposition that it is raining, and either snowing or blowing or both. The expression

$$(AB \vee C)$$

or (since the outermost parentheses are not needed to resolve ambiguity)

$$AB \vee C$$

can only be read as the disjunction of AB with C: it denotes the proposition that either it is both raining and snowing, or it is blowing, or it is doing all three. Parentheses provide a uniform method for resolving ambiguities: the methods provided by the English language are more various and complex.

When "*prob*" or "*des*" is written in front of a propositional expression, the result is a numerical expression which can appear in contexts similar to those in which the symbols

$$x, y, 0, 1, 2, \ldots$$

of ordinary arithmetic may appear. Thus a special case of the addition law for probabilities can be written

$$prob\ A + prob\ \overline{A} = 1;$$

the *general addition law* can be written

$$prob\ (A \vee B) = prob\ A + prob\ B - prob\ AB;$$

and the condition under which a proposition A is preferred to a proposition B can be written.

$$des\ A > des\ B.$$

The interpretation of such expressions and the rules for manipulating them will be discussed in the following chapter.

4.7 Belief versus Assent

To take the objects of belief and desire to be propositions rather than sentences is to rule out certain natural kinds of talk. If the objects of the two attitudes are sentences, it may be natural to say that the agent believes the sentence

It is not both raining and snowing

but does not believe the sentence

Either it is not raining or it is not snowing

even though these sentences are logically equivalent—even though they have the same meaning, so that the propositions that they express,

$$\overline{AB}$$

and

$$(\overline{A} \vee \overline{B}),$$

are one and the same. For somehow, the agent might fail to realize that the two sentences mean the same and might even be unable to follow a demonstration of that fact such as is given by comparing tables 4.2 (a) and (b) and thereby see that the two sentences at the right are either both true or both false in each of the four possible cases that are listed at the left. Thus, confronted with the two sentences, he might assent to one but fail to assent to the other, and we might therefore be inclined to say that he believes the first but not the second. And if we go beyond the elementary level of logic to which we are confining ourselves in this chapter, such situations are seen to be the rule, because, in general, there need be no routine way of testing pairs of sentences for logical equivalence, and therefore the possibility that the agent assents to one but not the other of a logically equivalent pair of sentences is entirely compatible with the possibility that he is a careful and competent logician.

Table 4.2

A	B	AB	$\overline{AB}$
true	*true*	*true*	*false*
true	*false*	*false*	*true*
false	*true*	*false*	*true*
false	*false*	*false*	*true*

(a)

A	B	$\bar{A}$	$\bar{B}$	$(\bar{A} \vee \bar{B})$
true	*true*	*false*	*false*	*false*
true	*false*	*false*	*true*	*true*
false	*true*	*true*	*false*	*true*
false	*false*	*true*	*true*	*true*

(b)

Similarly in the case of desire, it is conceivable that an agent might claim to desire the truth of the sentence

It is not both raining and snowing

but to be indifferent about the truth of the sentence

Either it is not raining or it is not snowing,

even though the truth conditions for the two sentences are precisely the same. Then, in taking propositions instead of sentences to be the objects of belief and desire, I seem to be assuming that the agent is a careful and competent logician; and if we consider sentences whose equivalence can only be demonstrated in more advanced branches of logic, it appears that I am assuming the agent to be a better logician than any man (or machine) could possibly be.

In rejecting these assumptions, one must also reject the overly simple account of the relation between belief in propositions and assent to sentences on which they are based. On that account of the matter, belief in a proposition implies willingness to assent to any sentence that expresses it. A more plausible account might say that belief in a proposition implies willingness to assent to at least one of the sentences that express it, for example, one of the shortest such sentences. If one then wanted to talk about belief in sentences, one could say that belief in a sentence implies willingness to assent to at least one sentence which is logically equivalent to it. Thus, the agent's belief in the sentence

Either it is not raining or it is not snowing

might be evidenced by his assent to another sentence

It is not both raining and snowing.

The agent would then be said to believe the first sentence even though he is unwilling to assent to it. I take the strangeness of this situation to be a symptom of the fact that propositions, not sentences, are the appropriate objects of belief. On the other hand (I claim), it seems natural enough to speak of willingness to assent to propositions, meaning thereby willingness to assent to some one of the sentences which express that proposition.

No doubt, we commonly apprehend propositions by means of sentences that express them, but for all that, we need not suppose that the agent would express his beliefs and desires by using the very sentences that we use in discussing his beliefs and desires. To say that Columbus believed that the earth is round is not to imply that he understood English. Nor need the agent be a user of any language: often enough, my cat's behavior makes it clear to me that he believes he is about to be fed, or that he desires to be fed, or that he prefers tuna to egg. In ordinary talk as in the theory of preference, we may find it convenient and natural to attribute certain propositional attitudes to the cat. In doing so we do not imply that the cat understands English, or is thinking of some proposition, or is thinking at all. The theory of deliberate action is ours, not the cat's, and the theory can be used to explain some of the cat's actions even though the cat does not understand the theory, just as the cat can digest his food without being a chemist. Of course, we do not discover what the cat's beliefs and desires are by putting questions to him, but then, questioning is only a rough and ready way of discovering the beliefs and desires of humans, who are often enough unwilling or unable to give the right answers. The criteria for belief and desire are behavioral, and speech is only a part of behavior.

4.8 Problems

1

Let H_1, H_2, and H_3 be the propositions that a certain coin lands head up the first, second, and third times it is tossed. Suppose the coin is tossed three times, and never lands on edge. Express as simply as you can in this notation the proposition that

(a) the first and second tosses both yield heads;
(b) the first and second tosses both yield tails;
(c) the first and second tosses have the same outcome;
(d) the first and second tosses have opposite outcomes;
(e) all tosses yield heads;
(f) there is at least one head;
(g) there is at least one tail;
(h) there are at least two heads;
(i) there are at most two heads;
(j) there are exactly two heads.

2

Table 4.2 shows that $\overline{AB}$ and $\overline{A} \vee \overline{B}$ are one and the same proposition: it shows that

$$\overline{AB} = \overline{A} \vee \overline{B} \, .$$

Use the same technique ("truth tables") to show that each of the following equalities or inequalities holds. Also, express each proposition by means of English sentences corresponding to the two sides of the equations, where A, B, C are the propositions that it is raining, it is snowing, it is blowing.

(a) $\overline{A \vee B} = \overline{A}\,\overline{B}$
(b) $A \vee \overline{A}B = A \vee B$
(c) $A(\overline{A} \vee B) = AB$
(d) $A(B \vee C) = AB \vee AC$ (These tables have
(e) $A \vee BC = (A \vee B)(A \vee C)$ eight rows.)
(f) $\overline{\overline{A}} = A$ (This table has two rows.)
(g) $A \vee \overline{A} = B \vee \overline{B}$
(h) $A\overline{A} = B\overline{B}$
(i) $\overline{AB} \neq \overline{A}\,\overline{B}$ (The propositions are distinct if the
(j) $A \vee B \neq \overline{A} \vee \overline{B}$ rightmost columns are not identical.)
(k) $AB \vee A\overline{B} \vee \overline{A}B = A \vee B$
(l) $AB \vee A\overline{B} \vee \overline{A}B \vee \overline{A}\,\overline{B} = A \vee \overline{A}$
(m) $AA = A$
(n) $A \vee A = A$
(o) $AB = BA$
(p) $A \vee B = B \vee A$
(q) $(AB)C = A(BC)$
(r) $(A \vee B) \vee C = A \vee (B \vee C)$

3

The probability of a proposition is the sum of the probabilities of the cases in which it is true. Thus if the probabilities of the four cases in table 4.2 (a) are

.1

.2

.3

.4

the probability of the proposition $\overline{AB}$ is

$$.2 + .3 + .4 = .9 \, .$$

Using the given probabilities for the four cases, find the probabilities of propositions (a), (b), (c), (g), (h) in problem 2.

4
Find the probabilities of propositions (a–j) in problem 1 on the assumption that all eight cases have the same probability .125.

5
Define

$$X \text{ implies } Y$$

as meaning that

$$XY = X$$

and show that:

(a) X implies Y if and only if $(X \vee Y) = Y$;
(b) every proposition implies itself;
(c) if X implies Y then $\overline{Y}$ implies $\overline{X}$ (law of contraposition);
(d) if X is true and X implies Y then Y is true (law of detachment);
(e) if X implies Y and Y implies Z then X implies Z (law of syllogism);
(f) if X implies Y and Y implies X then $X = Y$.

4.9 References and Solutions

The characterization of a proposition's probability as the value of the opportunity to gain a unit of value if the proposition is true certainly goes back to Bayes's 1763 essay (cited in section1.8) and can plausibly be read out of the first published monograph on probability, Christiaan Huygens, *De Ratiociniis in Ludo Aleae* [On calculating in games of luck], published in 1656–1657 (see Huygens's *Oevre Completes,* vol. 14, La Haye, 1920). For an illuminating account of the history of probability, see Ian Hacking, *The Emergence of Probability* (Cambridge: At the University Press, 1975). The "Dutch book" argument justifying the special addition law (section 4.2) is due to Frank Ramsey, "Truth and Probability," (cited in section 3.9) and, independently, to Bruno de Finetti, "La Prévision: ses lois logiques, ses sources subjectives." *Annales de l'Institut Henri Poincaré,* vol. 7, 1937. Both articles are reprinted, de Finetti's in English translation, in Kyburg and Smokler, *Studies in Subjective Probability* (cited in section 3.9), pp. 40–41 and 62–64.

Here are solutions to some of the problems in section 4.8.

Problem 1

(a) H_1H_2 (b) $\overline{H}_1\overline{H}_2$ (c) $H_1H_2 \vee \overline{H}_1\overline{H}_2$

(d) $H_1\overline{H}_2 \vee \overline{H}_1H_2$ (e) $H_1H_2H_3$ (f) $H_1 \vee H_2 \vee H_2$

(g) $\overline{H}_1 \vee \overline{H}_2 \vee \overline{H}_3$ (h) $H_1H_2 \vee H_2H_3 \vee H_1H_3$
(j) $\overline{H}_1H_2H_3 \vee H_1\overline{H}_2H_3 \vee H_1H_2\overline{H}_3$

Problem 2
(a), (b), (c), (g), (h). The rightmost columns of the tables should be as shown below:

A	B	(a)	(b)	(c)	(g)	(h)
t	t	f	t	t	t	f
t	f	f	t	f	t	f
f	t	f	t	f	t	f
f	f	t	f	f	t	f

Problem 3
The probabilities are .4, .6, .1, 1, 0.

Problem 5
(a) In table 4.3, the cases in which XY has the same truth value as X (cases 1, 3, 4) are exactly those in which $(X \vee Y)$ has the same truth value as Y.

Table 4.3

X	Y	XY	$X \vee Y$
t	t	t	t
t	f	f	t
f	t	f	t
f	f	f	f

(b) To show that an arbitrary proposition A implies itself, set $A = X$ and $A = Y$ in the definition of "implies" and compare with problem 2 (m).

5

Preference

The notions of numerical desirability and probability (relative to a certain agent at a certain time) will later be defined in terms of the agent's preference ranking of propositions. However, it is illuminating to begin by imagining that the notions of subjective probability and desirability are already available, and defining preference in terms of them. In this way, we shall see why preference, in the intended interpretation of that term, has the fundamental properties that we attribute to it and that we later employ to deduce the probability and desirability assignments from the preference ranking.

5.1 Computing Probabilities

For definiteness, imagine that we are concerned with the three propositions *A, B, C*:

A = that it will rain tomorrow
B = that it will snow tomorrow
C = that it will blow tomorrow

There are eight possibilities for the joint truth or falsity of these three propositions. The possibilities are enumerated at the left in table 5.1, where *ttt* is the case in which *A, B,* and *C* are all true, *ttf* is the case in which *A* and *B* are true but *C* is false, and so on down to *fff,* in which case *A, B,* and *C* are all false.

The probabilities that an agent attributes to the eight cases might be represented by any sequence of eight numbers, provided only that none of the numbers is negative, and that their sum is 1. In particular, the numbers listed under "*prob*" in table 5.1 might be the probabilities that someone attributes to the eight cases. The desirabilities that an agent attributes to the eight cases might be represented by any sequence of eight numbers, positive, negative, or zero, without restriction. In particular, the numbers listed under "*des*" in table 5.1 might be the desirabilities that someone attributes to the eight cases.

Table 5.1

	A	B	C	*prob*	*des*
1	*t*	*t*	*t*	.1	−2
2	*t*	*t*	*f*	.1	−1
3	*t*	*f*	*t*	.2	−1
4	*t*	*f*	*f*	.1	0
5	*f*	*t*	*t*	.2	−1
6	*f*	*t*	*f*	.1	0
7	*f*	*f*	*t*	.1	1
8	*f*	*f*	*f*	.1	2

Knowing the probabilities and desirabilities of the eight cases, one can compute the probabilities and desirabilities of all the possible propositions that can be built up out of the ingredients A, B, C, by using the operations of denial, conjunction, and disjunction. In the case of probabilities, the method is extremely simple: *the probability of a proposition is simply the sum of the probabilities of the cases in which it would be true*. Thus, since the proposition A is true in the first four cases, and false in all the others, the probability is the sum of the probabilities of the first four cases:

$$prob\ A = .1 + .1 + .2 + .1 = .5\ .$$

Since the proposition B is true in the first two cases and in cases 5 and 6, but false in all others, its probability is the sum of the probabilities of cases 1, 2, 5, 6:

$$prob\ B = .1 + .1 + .2 + .1 = .5\ .$$

And since C is true in the odd-numbered cases and false in all the rest, its probability is the sum of the probabilities of the odd-numbered cases:

$$prob\ C = .1 + .2 + .2 + .1 = .6\ .$$

Compound propositions are treated in the same way. Since $(A \vee B)$ is true in the first six cases only, its probability is the sum of the probabilities of the first six cases:

$$prob\ (A \vee B) = .8\ .$$

Since $\overline{C}$ is true in the even-numbered cases only, its probability is the sum of the probabilities of the even-numbered cases:

$$prob\ \overline{C} = .4\ .$$

And since $(A \vee \bar{A})$ is true in all eight cases, its probability is simply the sum of the probabilities of all eight cases:

$$prob\ (A \vee \bar{A}) = 1\ .$$

5.2 The Propositions *T* and *F*

There is just one necessary proposition. This is the proposition which is true in all cases, and is one and the same proposition whether it be denoted by the expression "$(A \vee \bar{A})$" or by the expression "$(B \vee \bar{B})$" or by the expression "$(C \vee \bar{C})$". Accordingly, it is convenient to have a special symbol

$$T$$

for this proposition, just as it is convenient to have the special symbol

$$0$$

for the number which is denoted by each of the expressions

$$x - x \quad y - y \quad z - z$$

in ordinary algebra. Since the necessary proposition T is true in all cases, and since the sum of the probabilities of all cases is 1, we have

$$prob\ T = 1\ .$$

Similarly, there is just one impossible proposition. This is the proposition which is denoted equally well by each of the expressions

$$A\bar{A} \quad B\bar{B} \quad C\bar{C}$$

—the proposition which is false in all cases. For the impossible proposition, we shall use the special symbol

$$F$$

and set

$$prob\ F = 0\ .$$

5.3 A Remark on Computing Probabilities

Given the probabilities of the eight cases shown in table 5.1, we can determine the probabilities of all propositions, including the probabilities of the propositions A, B, C to whose joint truth or falsity the eight cases refer. However, given only the probabilities of the three propositions A, B, C, we would not have been able to discover the probabilities of the eight cases. Thus, given the information that

$$prob\ A = prob\ B = .5\ , \quad prob\ C = .6\ ,$$

we would have known that

$$prob\ \overline{A} = prob\ \overline{B} = .5\ , \quad prob\ \overline{C} = .4\ ,$$

since the probability of the denial of a proposition is always 1 minus the probability of that proposition; and we would have known that

$$prob\ T = 1\ , \quad prob\ F = 0\ .$$

But we would not have known the value of

$$prob\ (A \vee B)\ ;$$

and concerning the eight cases, we would have known only the values of certain sums of their probabilities.

Example 1

Since *prob A* = .5 and since *A* is true in just the first four cases, we would know that the sum of the probabilities of the first four cases is .5:

$$p_1 + p_2 + p_3 + p_4 = .5\ ,$$

where the probability of case number n is p_n. Similarly, the fact that *prob B* = .5 would have told us the value of the sum of the probabilities of cases 1, 2, 5, 6:

$$p_1 + p_2 + p_5 + p_6 = .5\ .$$

Also, the fact that *prob C* = .6 would have told us the value of the sum of the probabilities of the odd-numbered cases:

$$p_1 + p_3 + p_5 + p_7 = .6\ .$$

To this information we can add the general truth that the sum of the probabilities of all cases is 1,

$$p_1 + p_2 + p_3 + p_4 + p_5 + p_6 + p_7 + p_8 = 1\ ,$$

and that none of the p's are negative. But this information is not enough to determine the values of the individual p's. Thus, the information given is compatible not only with the probability assignment given in table 5.1, but also with the following radically different assignment:

$$p_1 = .3$$

$$p_2 = 0$$

$$p_3 = .1$$

$$p_4 = .1$$

$$p_5 = .1$$

$$p_6 = .1$$

$$p_7 = .1$$

$$p_8 = .2$$

Relative to the probability assignment given in table 5.1, the probability of the proposition $A \vee B$ was .8; relative to this new probability assignment, the probability of that same proposition is .7. Now since both assignments are compatible with the conditions that the probabilities of A, B, C are to be .5, .5, .6, it is clear that the probabilities of A and B do not determine the probability of $A \vee B$. Rather, it is the probabilities of the eight possible cases that determine the probabilities of all compounds that can be built up by applying denial, conjunction, and disjunction to A, B, C.

5.4 Computing Desirabilities

Knowing the probabilities and the desirabilities of the eight cases, we can compute the desirabilities of all compounds of A, B, C, except for the impossible proposition F, to which we decline to attribute any desirability at all. (It is not that *des F* has the value 0; rather, *des F* has no value at all, not even the value 0. We leave the impossible proposition out of the preference ranking.)

The rule is this: *the desirability of a proposition is a weighted average of the desirabilities of the cases in which it is true, where the weights are proportional to the probabilities of the cases.* More explicitly, the procedure for computing the desirability of a proposition can be given in three steps: (1) multiply the desirability of each case in which the proposition is true by the probability of that case; (2) add all such products; and (3) divide by the probability of the proposition. (In other words, divide by the sum of the probabilities of the cases in which the proposition is true.)

Example 2

Use the probabilities and desirabilities given in table 5.1. Since A is true in just the first four cases, step (1) in the computation of *des A* results in the following four products:

$$(.1)(-2) = -.2$$

$$(.1)(-1) = -.1$$

$$(.2)(-1) = -.2$$

$$(.1)(0) = 0$$

Step (2) yields the sum

$$(-.2) + (-.1) + (-.2) + (0) = -.5 .$$

In step (3) we divide this sum by the probability of A, which is the sum,

$$.1 + .1 + .2 + .1 = .5 ,$$

of the cases in which A is true. The result is

$$des\ A = (-.5)/(.5) = -1 .$$

The entire computation could have been represented by a single formula,

$$des\ A = \frac{(.1)(-2) + (.1)(-1) + (.2)(-1) + (.1)(0)}{.1 + .1 + .2 + .1} = -1 .$$

Similarly, the calculation for $\overline{A}$, which is true in the last four cases, would have been

$$des\overline{A} = \frac{(.2)(-1) + (.1)(0) + (.1)(1) + (.1)(2)}{.2 + .1 + .1 + .1} = 1/5 .$$

The calculations for B, $\overline{B}$, C, $\overline{C}$ have the results

$$des\ B = -1 \quad des\ \overline{B} = 1/5 \quad des\ C = -5/6 \quad des\ \overline{C} = 1/4.$$

Since the necessary proposition is true in all eight cases, its desirability (combining some steps) is

$$des\ T = \frac{-.2 - .1 - .2 + 0 - .2 + 0 + .1 + .2}{1} = -.4 .$$

Then the preference ranking, as far as it concerns the propositions whose desirabilities we have computed, is as follows.

$\overline{C}$

$\overline{A}$, $\overline{B}$

T

C

A , B

We can expand this preference ranking by computing

$$des\ (A \vee B) = -7/8$$

and by noting that the eight cases *ttt, ttf, . . . , fff* whose desirabilities are given in table 5.1 are simply the propositions

$$ABC, AB\overline{C}, \ldots, \overline{A}\overline{B}\overline{C} .$$

A more complete preference ranking would be as follows, where the desirabilities of the ranks are indicated at the left.

2	$\overline{A}\overline{B}\overline{C}$
1	$\overline{A}\overline{B}C$
.25	$\overline{C}$
.2	$\overline{A}$, $\overline{B}$
0	$A\overline{B}\overline{C}$, $\overline{A}B\overline{C}$
$-.4$	T
$-.833$	C
$-.875$	$A \vee B$
-1	A, B, $AB\overline{C}$, $A\overline{B}C$, $\overline{A}BC$
-2	ABC

5.5 The Probability and Desirability Axioms

The rules for computing probabilities are implicit in the following three axioms.

(5-1) PROBABILITY AXIOMS

(a) prob is *nonnegative:* $prob\ X \geqslant 0$.
(b) prob is *normalized:* $prob\ T = 1$.
(c) prob is *additive:* if $XY = F$, then $prob\ (X \vee Y) = prob\ X + prob\ Y$

The first axiom asserts that probabilities cannot be negative; the second that the probability of the necessary proposition is 1; and the third, that if propositions X and Y are logically incompatible, then the probability that one or the other is true is simply the sum of their probabilities. Similarly, the rules for computing desirabilities are deducible from the probability axioms together with the following additional axiom.

(5-2) DESIRABILITY AXIOM. If $prob\ XY = 0$ and $prob\ (X \vee Y) \neq 0$, then

$$des\ (X \vee Y) = \frac{prob\ X\ des\ X + prob\ Y\ des\ Y}{prob\ X + prob\ Y} .$$

The desirability axiom says that the desirability of a disjunction $(X \vee Y)$ of incompatible propositions is a weighted average of the desirabilities of the incompatible ways (X, Y) in which it can come true, the weights (*prob* X, *prob* Y) being the probabilities of those ways.

It is instructive to deduce further laws about probabilities from the three probability axioms. These will be numbered consecutively with the axioms (5-1) (a), (b), (c).

(5-1) (d) $$prob\ \bar{A} = 1 - prob\ A\ .$$

Derivation. Since $A\bar{A} = F$, we can set $X = A$ and $Y = \bar{A}$ in (5-1) (c) to get $prob\ (A \vee \bar{A}) = prob\ A + prob\ \bar{A}$; and since $A \vee \bar{A} = T$, we can rewrite this as $prob\ T = prob\ A + prob\ \bar{A}$. Solving this equation for $prob\ \bar{A}$ and applying (5-1) (b) we then have (5-1) (d).

(5-1) (e) $$prob\ A\bar{B} = prob\ A - prob\ AB\ .$$

Derivation. Since $(AB)(A\bar{B}) = F$ we can set $X = AB$ and $Y = A\bar{B}$ in (5-1) (c) to get $prob\ (AB \vee A\bar{B}) = prob\ AB + prob\ A\bar{B}$. Since $AB \vee A\bar{B} = A$, we can rewrite this as $prob\ A = prob\ AB + prob\ A\bar{B}$, which can be solved for $prob\ A\bar{B}$ to get (5-1) (e).

(5-1) (f) $$prob\ A \geqslant prob\ AB\ .$$

Derivation. By (5-1) (a), probabilities cannot be negative. Therefore, the left-hand side of equation (5-1) (e) cannot be negative, and from this it follows that the first term on the right-hand side is at least as great as the second, and this is what (5-1) (f) states.

(5-1) (g) $$prob\ (A \vee B) = prob\ A + prob\ B - prob\ AB\ .$$

Since $A \vee B = B \vee A\bar{B}$, we have $prob\ (A \vee B) = prob\ (B \vee A\bar{B})$, and since $B(A\bar{B}) = F$, we then can apply (5-1) (c) to get $prob\ (A \vee B) = prob\ B + prob\ A\bar{B}$. Applying (5-1) (e) to the last term of this equation, we then have (5-1) (g).

The three probability axioms together with their consequences comprise the laws of the elementary *probability calculus*. (The term "calculus" is used here in its general sense, of a method of calculation.) Similarly, the elementary *desirability calculus* is obtained by adding the desirability axiom (5-2) to the three probability axioms. These four axioms together with their consequences yield the same results as the rule that was given earlier (in English) for computing desirabilities of propositions as weighted averages of the desirabilities of the cases in which they would be true.

5.6 "Good," "Bad," "Indifferent"

We shall have frequent use for the following law of the desirability calculus, which is obtained by setting $Y = \bar{X}$ in (5-2).

(5-3) $$des\ T = prob\ X\ des\ X + prob\ \bar{X}\ des\ \bar{X}$$

The desirability of the necessary proposition is a weighted average of the desirabilities of X and of $\overline{X}$, in which the weights are the probabilities of X and of $\overline{X}$. Since this is true for any proposition X, the necessary proposition must occur with or between each proposition and its denial in the preference ranking. In particular, if *prob* X is 1, the necessary proposition will be ranked with X; if *prob* X is 0, the necessary proposition will be ranked with $\overline{X}$: and if *prob* X is neither 0 nor 1, the necessary proposition will be ranked strictly between X and $\overline{X}$ (unless X and $\overline{X}$ are ranked together). Suppose, in particular, that the preference ranking is

$$A$$
$$T$$
$$\overline{A}$$

Since T is $A \vee \overline{A}$, you can think of the necessary proposition as a gamble on A, in which the "gain" is

$$des\ A - des\ T$$

units of desirability if you win, and the "loss" is

$$des\ T - des\ \overline{A}$$

units of desirability if you lose.

Since T comes somewhere at or between A and $\overline{A}$ in the preference ranking, it is appropriate to say that A is *good, bad,* or *indifferent* accordingly as A is ranked *above, below,* or *with* T. In terms of desirabilities, A is good if $des\ A > des\ T$; bad if $des\ A < des\ T$; and indifferent if $des\ A = des\ T$. If we set

$$des\ T = 0\ ,$$

the situation is even simpler: A is good, bad, or indifferent accordingly as its desirability is positive, negative, or zero.

5.7 Preference between News Items

In general, to say that

$$des\ A > des\ B$$

is to say that A is ranked higher than B in the agent's preference ranking of propositions. But what does that mean? One way of looking at the matter, suggested to me by Savage, is in terms of preference between news items. To say that A is ranked higher than B means that the agent would welcome the news that A is true more than he would the news that B is true: A would be better news than B. This explains the fact that the impossible proposition does not occur in any preference ranking of propositions, because there could be no such news as

> It will rain all day tomorrow in San Francisco and it will not rain all day tomorrow in San Francisco,

and accordingly, it is idle to ask whether *A* is better news than *F*. In another sense it is not news that

> It will rain all day tomorrow in San Francisco or it will not rain all day tomorrow in San Francisco.

It is no news that *T* is true, not because *T* cannot be true, but because it must be true: the agent knew it all along. Then to say that *A* is ranked lower than *T* means that *A* is bad news in the sense that no news is good news, compared with the news that *A* is true; to say that *A* is ranked above *T* means that no news is bad news, compared with the (good) news that *A* is true; and to say that *A* is ranked with *T* means that the agent is indifferent to the news that *A* is true: he would be pleased no more and no less by the news that *A* is true than he would be by no news at all.

5.8 Acts as Propositions

It might be objected that preference, interpreted in this way, has little relevance to action: the agent is construed passively, as reacting with cheers, groans, or yawns to news items, or as greeting one hypothetical news item more or less enthusiastically than another, or greeting them with equal fervor. To answer this objection, notice that acts can often be described at least approximately by declarative sentences. Thus, in examples that we have considered, some of the acts are described by the following sentences.

> We have red wine with dinner.
> The agent takes the plane.
> The agent disarms.

(In the last case the agent is presumably a state.) These characterizations are rough, but perhaps accurate enough in relation to the relevant deliberations. The roughness shows itself in many ways, one pervasive mode of roughness being that one can always take a strict point of view from which the agent can only try to perform the act indicated by the declarative sentence. He can try to bring red wine, but may fail through picking up the wrong bottle, or by dropping the bottle en route. He can try to take the plane, but may fail through delay in traffic or through failure of the plane to take off. For the case where the agent is a composite social entity such as a state, a navy, a corporation, a club, a family, or a symphony orchestra, the term "try" will have one or another sense, depending on the nature and organization of the composite entity, and on the act which is contemplated.

However, where acts are characterized with sufficient accuracy by declarative sentences, we can conveniently identify the acts with the propositions that the sentences express. An act is then a proposition which is within the agent's power to make true if he pleases, and the necessary proposition would correspond to not acting: to letting what will be, be. But deliberate inaction is a form of action in an attenuated but useful sense of that term, and accordingly we might think of T as one of the available acts; certainly, noninterference with the status quo may be one of the courses open to the agent, and to call that course a course of action is not to stretch meanings beyond ordinary usage.

Now, to the extent that acts can realistically be identified with propositions, the present notion of preference is active as well as passive: it relates to acts as well as to news items. If R (red wine with dinner) and W (white wine with dinner) are construed as propositions, and if they are each within the agent's power to bring about if he sees fit, then for R to be ranked above W is for the agent to think the act of bringing red wine preferable to the act of bringing white. From this viewpoint, the notion of preference between propositions is neutral, regarding the active-passive distinction. If the agent is deliberating about performing act A or act B, and if AB is impossible, there is no effective difference between asking whether he prefers A to B as a news item or as an act, for he makes the news.

Example 3: Probabilities of acts

In example 3 of chapter 1, the desirability matrix for the guest's deliberation was given as

	Chicken	Beef
White	1	−1
Red	0	1

Construing acts and conditions as propositions, and using the assumptions that $B = \overline{C}$ (beef certainly will be served if chicken is not, and will not be served if chicken is), and that $R = \overline{W}$ (the wine will be red or white, one but not both), we have

$$des\ WC = 1 \qquad des\ W\overline{C} = -1$$

$$des\ \overline{W}C = 0 \qquad des\ \overline{W}\overline{C} = 1$$

At the end of example 4 in chapter 1, the probability matrix was given as

.75	.25
.25	.75

These are conditional probabilities: the probabilities of chicken and beef are .75, .25 if white wine is served, while the probabilities are .25, .75 if not. If the probability that white wine is served is p, then we have the following absolute probabilities:

$$prob\ WC = .75p \qquad prob\ W\overline{C} = .25p$$

$$prob\ \overline{W}C = .25(1 - p) \qquad prob\ \overline{W}\overline{C} = .75(1 - p)$$

Now we can compute the desirabilities of W, C, and all of their compounds in terms of the parameter p. For the acts W and $\overline{W}$, the parameter cancels out, yielding definite values

$$des\ W = \frac{(.75p)(1) + (.25p)(-1)}{p} = .5$$

$$des\ \overline{W} = \frac{.25(1 - p)(0) + .75(1 - p)(1)}{1 - p} = .75\ .$$

Then the act $\overline{W}$ of bringing red wine is preferred. If W were totally within the agent's power to make happen or not, we should then have $prob\ \overline{W} = 1$, since the preferred act will be performed. Then p is 0, which would seem to vitiate the calculation of $des\ W$, in which we divided by p. However, the situation can be saved by adopting the strict point of view, in which the agent can only try to make W true, or to make W false. We then reflect that if p is taken to be between 0 and 1 the calculation is legitimate, and that for all such values of p (be they ever so close to 0), the outcome is the same: $des\ \overline{W} = .75$, $des\ W = .5$.

5.9 Desirabilities Determine Probabilities

The fact that T can be expressed as $(X \vee \overline{X})$ for any proposition X establishes a close connection between the probabilities and the desirabilities of propositions that can come true in more than one possible case. It remains true that, as indicated in table 5.1, assignments of probabilities and desirabilities to the basic cases can be made quite independently of each other. Still, writing $prob\ \overline{X}$ as $1 - prob\ X$ in equation (5-3), we obtain

$$des\ T = (prob\ X)(des\ X) + des\ \overline{X} - (prob\ X)(des\ \overline{X})\ .$$

This equation can be solved for $prob\ X$ if $des\ X \neq des\ \overline{X}$:

(5-4) $$prob\ X = \frac{des\ T - des\ \overline{X}}{des\ X - des\ \overline{X}}.$$

Then the desirability assignment determines the probability assignment! The result can be made to appear even more startling if we set

$$des\ T = 0\ ,$$

in which case we have

$$prob\ X = \frac{-des\ \overline{X}}{des\ X - des\ \overline{X}}$$

if $des\ X \neq des\ \overline{X}$. Dividing the numerator and denominator by $-des\ \overline{X}$, we then obtain

(5-5) $$prob\ X = \frac{1}{1 - (des\ X / des\ \overline{X})}$$

provided $des\ X \neq des\ \overline{X}$ and $des\ T = 0$. Thus, the probability of a proposition is determined by the desirabilities of that proposition and its denial. The reason is, of course, that a proposition which would be true in more than one case is, in effect, a gamble between the cases in which it would be true. Therefore, in computing the desirability of such a proposition, we had to make use of the probabilities of its truth cases. Equation (5-5) exploits this fact and uses it in reverse to recover the probability of a proposition from its desirability and the desirability of its denial. We could just as well have solved equation (5-3) in such a way as to express the desirability of $\overline{X}$ in terms of the probability of X and the desirability of X: if $des\ T = 0$ and $des\ X \neq 1$, we have

(5-6) $$des\ \overline{X} = -(prob\ X / prob\ \overline{X})(des\ X)\ .$$

Thus, the desirability of a proposition's failing is obtained by changing the sign of the desirability of its holding and multiplying by the ratio of the probabilities that the proposition holds and fails—if $des\ T = 0$.

Example 4: Thermonuclear war

As indicated in chapter 2, it is to be expected that two desirability assignments can be made arbitrarily (provided the preferred proposition is assigned the larger desirability) without prejudicing any deliberation. It is convenient to use one of the arbitrary choices to set

$$des\ T = 0\ .$$

Suppose that we use the other choice to set

$$des\ \overline{W} = 1\ ,$$

where $\overline{W}$ is the proposition that there will be no thermonuclear war in this century, and suppose that the agent thinks the probability of $\overline{W}$ is .9. Then, setting $X = \overline{W}$ in equation (5-6), he must attribute desirability

$$des\ \overline{\overline{W}} = des\ W = -(.9/.1)(1) = -9$$

to the prospect of a thermonuclear war in this century. In effect, he is viewing the necessary proposition as a gamble on $\overline{W}$ in which he gets 1 unit of desirability if he wins, and *des* W units if he loses. Since he takes the probability of winning to be .9, and the expected desirability of the gamble to be 0, the desirability of W must be -9.

The dependence of probability on desirability is a result of the fact that when a proposition can come true in more than one way, it is, in effect, a gamble. Now the distinction between desirabilities and estimated desirabilities become useless when we consider that any analysis into mutually exclusive, collectively exhaustive cases can be refined by taking a further proposition into account. Relative to a single proposition A, the cases are

$$A, \overline{A}\ .$$

But nothing prevents us from making use of the fact that each of these cases can in turn be split into two: A becomes two cases,

$$AB, A\overline{B}\ ,$$

and $\overline{A}$ becomes two cases,

$$\overline{A}B, \overline{A}\overline{B}$$

And each of these four cases can in turn be split into two by considering a third proposition, C, etc. Then the distinction between propositions that can be true in only one way and those that are gambles is relative to the number of cases that we see fit to consider; and therewith the distinction between desirability and estimated desirability vanishes. In the future, we shall speak simply of the desirabilities of propositions, understanding that all desirabilities are to be viewed as estimated desirabilities.

5.10 Problems

1

Find the preference ranking of the four propositions $A, \overline{A}, B, \overline{B}$ when the probabilities and desirabilities of the basic cases are as follows.

A	B	*prob*	*des*
t	t	.4	1
t	f	.3	2
f	t	.2	3
f	f	.1	4

2

Which of the following preference rankings are impossible?

(a)	(b)	(c)	(d)	(e)
$A, \overline{A}$	B	$A, \overline{A}$	A	A
$B, \overline{B}$	$A, \overline{A}$	B	$\overline{B}$	$\overline{A}$
	$\overline{B}$	$\overline{B}$	$\overline{A}$	B
			B	$\overline{B}$

3

The agent believes it is more likely to rain than to snow tomorrow; thinks that rain or snow tomorrow would be bad; and, in fact, thinks that rain tomorrow would be just as bad as snow tomorrow. Is the proposition that

it will not rain tomorrow

with, above, or below the proposition that

it will not snow tomorrow

in his preference ranking?

4

The probability of rain tomorrow is .75; the probability of snow tomorrow is .50; and the probability of rain and snow tomorrow is .25. What can you conclude about the probability of

(a) rain and/or snow tomorrow?
(b) rain but no snow tomorrow?
(c) rain or snow, but not both, tomorrow?

5

How many different propositions are there in list (a)? In list (b)? In list (c)?

(a) $\overline{RS}$; $\overline{R}\,\overline{S}$; $\overline{R \vee S}$; $\overline{R} \vee \overline{S}$
(b) RR; $R\overline{R}$; $R \vee R$; $R \vee \overline{R}$; $S\overline{S}$; $S \vee \overline{S}$; R; T
(c) $R\overline{RS}$; $R \vee S$

6

Find the probability of the proposition R in case the desirability scale is as shown in (a); in (b); in (c); in (d).

(a)	(b)	(c)	(d)
R	R	$\overline{R}$	$\overline{R}$
	T		T
T		T	R
$\overline{R}$	$\overline{R}$	R	

7

Excluding the impossible proposition, there are 15 distinct propositions that can be compounded out of the 2 propositions A, B (counting A and B themselves among the 15). List these, and write out their preference ranking, based on the following assignment of probabilities and desirabilities to the basic cases.

A	B	*prob*	*des*
t	*t*	1/3	1
t	*f*	1/3	2
f	*t*	1/3	3
f	*f*	0	4

8

If AB is impossible and *prob* $(A \vee B) \neq 0$, then $A \vee B$ must be ranked with or between A and B. However, if AB is possible, $A \vee B$ may be better than either A or B alone, or worse. Thus, the following preference ranking might arise where A is the proposition that it will rain and B is the proposition that it will snow:

$$A \vee B$$

$$A, B$$

Assign probabilities and desirabilities to the four basic cases in such a way as to obtain this preference ranking.

9 Death before Dishonor

One might suppose that the operation of denial reverses preference rankings: that if A is preferred to B, $\overline{B}$ must be preferred to $\overline{A}$. To see that this need not be so, imagine that the agent is a Roman matron who prefers death to dishonor, and for whom dishonor entails death. (She would commit suicide if dishonored.) Explicitly, A is the proposition that the matron is dead by next week, B is the proposition that she is dishonored this week, $\overline{A}B$ has probability 0, and she prefers A to B. Explain why she also prefers $\overline{A}$ to $\overline{B}$.

10

Show that if *des* $T = 0$ then a good or bad proposition A has probability 1/2 if and only if *des* $\overline{A} = -$*des* A.

11

Show that if A and B are ranked together but not with T, then *prob* $A =$ *prob* B if and only if $\overline{A}$ and $\overline{B}$ are ranked together.

12

As in problem 5 of section 4.8, define "X implies Y" as meaning that $XY = X$. Show that

(a) F implies every proposition.
(b) Every proposition implies T.
(c) If X implies Y, then *prob* $X \leqslant$ *prob* Y.

13 Conditional probability

Suppose that your beliefs and desires change from those characterized by the pair *prob, des* to those characterized by a new pair, *PROB, DES,* simply because you have come to fully believe a proposition E that you had not fully believed or fully disbelieved before. Under such circumstances it is plausible to suppose that your new valuation of each proposition X will simply be your old valuation of XE:

$$\text{(5-7)} \qquad DES\ X = des\ XE$$

Use (5-4) and the probability and desirability laws of section 5.5 to deduce

$$\text{(5-8)} \qquad PROB\ H = \frac{prob\ HE}{prob\ E}$$

from the foregoing assumptions, on the hypothesis that HE and $\overline{H}E$ were not ranked together in your old (*des*) preference ordering. (This problem is solved at the end of section 5.11.)

14

Show that in the presence of (5-1) (a, b), (5-1) (c) implies

$$\text{(5-1)(h)} \quad \text{if } prob\ XY = 0, \text{ then prob } (X \vee Y) = prob\ X + prob\ Y$$

but not vice versa. (Hint: what if *prob* assigned the value 1 to all propositions?) Show that in the presence of (5-1), (5-2) implies and is implied by

(5-9) If $XY = F$ and $prob\ (X \vee Y) \neq 0$, then

$$des(X \vee Y) = \frac{prob\ X\ des\ X + prob\ Y\ des\ Y}{prob\ X + prob\ Y}.$$

15

The agent's preference ranking is as follows, where C, S, and R are the propositions that *a club is drawn, a spade is drawn,* and *a red card is drawn* from a certain standard deck.

C

$S, C \vee R$

R

If $des\ C = 1$, $des\ T = 0$, and if the agent attributes the usual probabilities to the propositions C, S, R, what are (a) *des S?* (b) *des R?*

16

Suppose that A and B are pairwise incompatible propositions, and suppose that the preference ranking is as follows.

A, B, G

T

$\overline{G}, \overline{A \vee B}$

$\overline{A}, \overline{B}$

Suppose that $des\ G = 1$ and $des\ T = 0$.

(a) Assuming that $prob\ G = 1/2$, find $prob\ A$, $prob\ B$, $des\ \overline{G}$, and $des\ \overline{A}$.
(b) Assuming that $des\ \overline{G} = -2$, find $des\ \overline{A}$, $prob\ A$, $prob\ B$, $prob\ G$.

5.11 Notes and References

The necessity discussed in section 5.2 is logical, not causal. There is only one logically necessary proposition, *T*, and only one logically impossible proposition, *F*. Equally important, but very different, is the notion of necessity discussed by Aristotle in *De Interpretatione*, chapter 9:

> What is, necessarily is, when it is . . . I mean, for example: it is necessary for there to be or not to be a sea-battle tomorrow; but it is not necessary for a sea-battle to take place tomorrow, nor for one not to take place—though it is necessary for one to take place or not to take place. [From *Aristotle's "Categories" and "De*

Interpretatione,'' trans. John Ackrill (Oxford: Oxford University Press, 1969).]

In ''Aristotle and the Sea Battle,'' *Mind* 65 (1956): 1–15, G. E. M. Anscombe explains:

> Aristotle's point . . . is that ''Either p or not p'' is always necessary, and this necessity is what we are familiar with. But . . . that when p describes a present or past situation, then either p is necessarily true, or $\bar{p}$ is necessarily true; and here ''necessarily true'' has a sense which is unfamiliar to us. In this sense I say it is necessarily true . . . that there is a University in Oxford; and so on. But ''necessarily true'' is not simply the same as ''true''; for while it may be true that there will be rain tomorrow, it is not necessarily true. As everyone would say: there may be or may not.

(I have replaced Anscombe's ''$\sim p$'' by our notation ''$\bar{p}$'' for ''not-p.'') For more about this, and further references, see my paper, ''Coming True,'' in *Intention and Intentionality,* ed. Cora Diamond and Jenny Teichman (Brighton: Harvester Press, 1979).

Example 3 (Probabilities of acts) is important. Put it like this: in deliberating about whether to act so as to make W true or so as to make it false, it would be idle to try to determine one's current degree of belief in W (say, by ''introspection''), for one knows that at the end of deliberation one will be in possession of a much more useful degree of belief in W. Call it ''p,'' i.e., the degree of belief (near 1 or 0) one will have upon deciding to act so as to make W true, or so as to make it false. As we noticed in section 1.7, the wise agent will realize that he may not actually manage to perform his chosen act: p will not be quite 1, or quite 0. Normally, e.g., in cases like example 3, in coming to have p as his degree of belief in W, i.e., in coming to choose, the agent does not come to alter his (conditional) probability matrix. But the point of example 3 does not depend on its being ''normal'' in that way. (I go on about this in an excessively technical way in ''A Note on the Kinematics of Preference,'' *Erkenntnis* **11** [1977]: 135–41.)

About problem 9. A less extreme example in which denial fails to reverse preferences arises as a counterexample to Aristotle's claim that

> A is preferable to B, if A is an object of choice without B, while B is not an object of choice without A; for example, power is not an object of choice without prudence, but prudence is an object of choice without power. [*Topics* III, Loeb Classical Library (1960), 118a 20.]

In our terms, Aristotle's claim is that the following inference is valid:

A is choiceworthy without B.	$des\ A\overline{B} \geqslant des\ \overline{A}\overline{B}$
B is not choiceworthy without A.	$des\ B\overline{A} < des\ \overline{B}\overline{A}$
A is more choiceworthy than B.	$des\ A \quad > des\ B$

Note that the premises imply that $\overline{B}$ (in the interval from $\overline{A}\overline{B}$ to $A\overline{B}$) is preferred to $\overline{A}$ (in the interval from $B\overline{A}$ to $\overline{B}\overline{A}$, i.e., the interval from $\overline{A}B$ to $\overline{A}\overline{B}$). But with these desirabilities and probabilities for the four conjunctions

	$\overline{A}B$	$\overline{A}\overline{B}$	$A\overline{B}$	AB
des	0	10	20	40
prob	.05	.05	.45	.45

we can compute the desirabilities of A and B as 30 and 36, contrary to Aristotle's conclusion. Aristotle's conclusion follows if we make the additional assumption that the four conjunctions are equiprobable (see *Synthese* 48 [1981]: 743–45).

Solution to problem 13. By (5-4), (5-7), and the fact that $TE = E$,

$$PROB\ H = \frac{DES\ T - DES\ \overline{H}}{DES\ H - DES\ \overline{H}} = \frac{des\ E - des\ \overline{H}E}{des\ HE - des\ \overline{H}E}.$$

By (5-2) and (5-1) (c), we may set

$$des\ E = \frac{des\ HE\ prob\ HE + des\ \overline{H}E\ prob\ \overline{H}E}{prob\ E}$$

in this, to get

$$PROB\ H = \frac{des\ HE\ prob\ HE + des\ \overline{H}E\ prob\ \overline{H}E - des\ \overline{H}E\ prob\ E}{prob\ E\ (des\ HE - des\ \overline{H}E)}$$

or, applying (5-1) (e) in the form $prob\ \overline{H}E = prob\ E - prob\ HE$ to the middle term of the numerator and then cancelling and factoring, we have (5-8):

$$PROB\ H = \frac{prob\ HE\ (des\ HE - des\ \overline{H}E)}{prob\ E\ (des\ HE - des\ \overline{H}E)} = \frac{prob\ HE}{prob\ E}.$$

Observe that the assumption (5-7) plays an essential role in this derivation, which is thus no proof that whenever you come to have full belief in E, your beliefs in other propositions X should change from $prob\ X$ to $prob\ XE/prob\ E$. Indeed, there can be no such general proof, for the claim is false (see chapter 11). The ratio $prob\ HE/prob\ E$ is abbreviated $prob(H/E)$ and called "the conditional probability of H on E." (Here it is assumed that *prob E* is positive, so that the ratio does not have the indeterminate form 0/0.) The business of changing beliefs from those characterized by *prob* to those characterized by *prob* (/*E*) is called "conditionalization" or "conditioning" on *E*.

6

Equivalence, Perspectives, Quantization

In Ramsey's theory, an agent's preference ranking of consequences and gambles between consequences is represented by a pair of assignments

$$prob,\ des\ ,$$

where *prob* assigns numerical probabilities to the propositions on which the agent may gamble, and *des* assigns numerical desirabilities to the consequences that may be the results of the gambles, and to the gambles themselves. The probability assignment must satisfy the laws of the elementary probability calculus which are embodied in equation (5-1) (a), (b), and (c); and the desirability assignment must satisfy the additional requirement that the (estimated) desirability of a gamble

$$B \text{ if } A,\ C \text{ if not}$$

must be

$$prob\, A\ des\, AB + prob\, \overline{A}\ des\, \overline{A}C\ .$$

If the preference ranking is sufficiently extensive, the probabilities are then uniquely determined by it, and the desirabilities are uniquely determined by it once two desirabilities have been assigned arbitrarily. In particular, suppose that

$$prob,\ des$$

and

$$PROB,\ DES$$

are two pairs of assignments, both of which represent a certain preference ranking. Ramsey was able to show that if the preference ranking is sufficiently extensive, the two probability assignments must be identical, so that for any proposition X on which the agent may gamble, we have

$$PROB\ X = prob\ X\ ,$$

and one desirability assignment can be converted into the other by choosing a suitable positive number a and a number b and applying the equation

$$DES\ X = a\ des\ X + b\ ,$$

where X is any consequence or gamble. Furthermore, if the pair *prob, des* represents a preference ranking, we can choose any positive number a, and any number b, and apply the foregoing equations to get a pair *PROB, DES* which also represents the given preference ranking.

6.1 Bolker's Equivalence Theorem

In the present theory, the role of the foregoing equations is played by two rather more complicated equations. In place of the two constants a, b, we have four constants a, b, c, d; the new probability assignment need not be identical to the old, but instead may be related to it by an equation of form

$$\text{(6-1)} \qquad PROB\ X = (prob\ X)(c\ des\ X + d)\ ,$$

and the new desirability assignment will be related to the old by an equation of form

$$\text{(6-2)} \qquad DES\ X = \frac{a\ des\ X + b}{c\ des\ X + d}$$

(where a and c need not be positive. Now we shall describe this situation precisely and prove that the description is correct. The first part of the description is given by the following *equivalence theorem;* the remaining part is given by the *uniqueness theorem* of chapter 8.

A pair, *prob, des,* will be said to meet the *existence condition,* relative to a given preference ranking of propositions, if the probability and desirability axioms (5-1) and (5-2) are satisfied and the second member of the pair mirrors the preference ranking in the sense that *des X* is at least as great as *des Y* whenever X is ranked at least as high as Y. Similarly, a different pair, *PROB* and *DES,* will be said to meet the existence condition relative to the same preference ranking if the probability and desirability axioms are again satisfied (with "*prob*" and "*des*" now capitalized in [5-1] and [5-2]), and if *DES* also mirrors the preference ranking in the sense that *DES X* is at least as great as *DES Y* whenever X is ranked at least as high as Y. Now the equivalence theorem can be stated.

EQUIVALENCE THEOREM (Ethan Bolker, 1964). Suppose that the existence condition is met by a pair *prob, des,* that equations (6-1) and (6-2) hold for all X in the preference ranking, and that conditions (6-3) below are satisfied. Then the pair *PROB, DES* also meets the existence condition.

(6-3) (a) $ad - bc$ is positive.
(b) For each X in the preference ranking, $c\ des\ X + d$ is positive.
(c) $c\ des\ T + d = 1$.

To prove the equivalence theorem, it is sufficient to show that if (6-1) through (6-3) are all satisfied and the pair *prob, des* meets the existence condidion, then the following five requirements are met, and therefore the pair *PROB, DES* also meets the existence condition:

(i) The values of *PROB* are never negative;
(ii) $PROB\ T = 1$;
(iii) $PROB\ (X \vee Y) = PROB\ X + PROB\ Y$ if $XY = F$;
(iv) if $PROB\ XY = 0$ but $PROB\ (X \vee Y) \neq 0$, then

$$\mathrm{DES}(X \vee Y) = \frac{PROB\ X\ DES\ X + PROB\ Y\ DES\ Y}{PROB\ X + PROB\ Y}$$

(v) $DES\ X \leq DES\ Y$ if and only if $des\ X \leq des\ Y$.

In the proof we shall make much use of equations (5-1) (a), (b), (c), (5-2), and their consequences. That these equations hold is implied by the hypothesis that the pair *prob, des* meets the existence condition.

To verify (i), apply (6-3) (b), (6-1), and the fact that $prob\ X$ cannot be negative.

To verify (ii), set $X = T$ in (6-1) and apply (6-3) (c) and the fact that $prob\ T = 1$.

To verify (iii), consider two cases. Case 1: $prob\ (X \vee Y) = 0$. Then $prob\ X = prob\ Y = 0$ so that by (6-1), (iii) is $0 = 0 + 0$. Case 2: $prob\ (X \vee Y) \neq 0$. By (6-1) we have

$$PROB\ (X \vee Y) = prob\ (X \vee Y)\ [c\ des\ (X \vee Y) + d]\ ,$$

or, applying Equation (5-2) as we may, since $XY = F$,

$$\begin{aligned} PROB\ (X \vee Y) &= prob\ (X \vee Y) \left[c\ \frac{prob\ X\ des\ X + prob\ Y\ des\ Y}{prob\ (X \vee Y)} + d \right] \\ &= c\ prob\ X\ des\ X + c\ prob\ Y\ des\ Y + d\ prob\ X \\ &\quad + d\ prob\ Y \\ &= (prob\ X)(c\ des\ X + d) + (prob\ Y)(c\ des\ Y + d) \\ &= PROB\ X + PROB\ Y\ . \end{aligned}$$

To verify (iv), assume that $PROB\ XY = 0$ and that $Prob\ (X \vee Y) \neq 0$. By (6-1) we have

$$(prob\ XY)(c\ des\ XY + d) = 0\ ,$$

so that by (6-3) (b) with XY for X, $prob\ XY = 0$. Also by (6-1) we have

$$prob\ (X \vee Y)(c\ des\ (X \vee Y) + d) \neq 0\ ,$$

so, in particular, $prob\ (X \vee Y) \neq 0$. Now by (6-2),

$$DES(X \vee Y) = \frac{a\ des\ (X \vee Y) + b}{c\ des\ (X \vee Y) + d}$$

Since $prob\ XY = 0$ and $prob\ (X \vee Y) \neq 0$, we may apply equation (5-2) in the numerator and denominator on the right-hand side of this last equation to get

$$DES\ (X \vee Y) = \frac{a \dfrac{prob\ X\ des\ X + prob\ Y\ des\ Y}{prob\ X + prob\ Y} + b}{c \dfrac{prob\ X\ des\ X + prob\ Y\ des\ Y}{prob\ X + prob\ Y} + d}\ .$$

Simplifying and collecting terms, this becomes

$$DES\ (X \vee Y) = \frac{(prob\ X)(a\ des\ X + b) + (prob\ Y)(a\ des\ Y + b)}{(prob\ X)(c\ des\ X + d) + (prob\ Y)(c\ des\ Y + d)}\ .$$

After multiplying the respective terms of the numerator by

$$\frac{c\ des\ X + d}{c\ des\ X + d} \quad \text{and} \quad \frac{c\ des\ Y + d}{c\ des\ Y + d}\ ,$$

we can apply (6-2) and (6-1) to get

$$DES\ (X \vee Y) = \frac{PROB\ X\ DES\ X + PROB\ Y\ DES\ Y}{PROB\ X + PROB\ Y}$$

as required.

Finally, to verify (v), notice that by (6-2) we have $DES\ X \leqslant DES\ Y$ if and only if

$$\frac{a\ des\ X + b}{c\ des\ X + d} \leqslant \frac{a\ des\ Y + b}{c\ des\ Y + d}\ .$$

When both sides of an inequality are multiplied by the same *positive* number the new inequality is equivalent to the old: it is true if and only if the old inequality was true. Now by (6-3) (b), we may cross-multiply to get an equivalent inequality,

$$ac\,(des\ X)(des\ Y) + ad(des\ X) + bc(des\ Y) + bd$$
$$\leqslant ac(des\ X)(des\ Y) + bc(des\ X) + ad(des\ Y) + bd\ ,$$

or, equivalently,

$$(ad - bc)(des\ X) \leqslant (ad - bc)(des\ Y)\ .$$

By (6-3) (a), we can now multiply both sides by $1/(ad - bc)$ to find that the inequality

$$des\ X \leqslant des\ Y$$

is equivalent to the original inequality,

$$DES\ X \leqslant DES\ Y\ .$$

This completes the proof of the equivalence theorem.

6.2 Zero and Unit

The surprising feature of transformation (6-2) is that even if we have

$$DES\ A = des\ A \qquad DES\ B = des\ B\ ,$$

where A and B are not ranked together, it need not be the case that *des* and *DES* agree as to the desirabilities they assign to all propositions. To ensure total agreement, it is necessary that we also have

$$DES\ C = des\ C$$

for some C that is ranked neither with A nor with B. Throughout the remainder of this chapter we shall focus attention on this unexpected feature by supposing that *DES* and *des* agree as to the desirabilities they assign to two differently ranked propositions. For definiteness, we shall suppose that

(6-4) $$DES\ T = des\ T = 0 \qquad DES\ G = des\ G = 1\ .$$

(The choice of these particular propositions and these particular values is a matter of convenience only.) Then we can set $X = T$ in (6-1) to get

$$1 = (1)(c.0 + d)$$

or

$$d = 1\ .$$

We can set $X = T$ in (6-2) to get

$$0 = \frac{a.0 + b}{c.0 + d}\ ,$$

or, since the denominator is positive,

$$b = 0 .$$

And, finally, we can set $X = G$ in (6-2) to obtain

$$1 = \frac{a.1 + 0}{c.1 + 1}$$

or

$$a = c + 1 .$$

Then equations (6-3) (b), (6-2), and (6-1) assume the forms

$$c \; des\; X > -1 \tag{6-5}$$

$$DES\; X = \frac{(c + 1) des\; X}{c\; des\; X + 1} \tag{6-6}$$

$$PROB\; X = (prob\; X)(c\; des\; X + 1) \tag{6-7}$$

when assumption (6-4) holds. It is worth noticing that, setting $X = G$ in (6-7), we then have

$$c + 1 = a = \frac{PROB\; G}{prob\; G} . \tag{6-8}$$

6.3 Bounds on Desirabilities

Then even when (6-4) holds, *PROB* need not agree with *prob,* nor need *DES* agree with *des,* on the values they assign to all propositions. However, if there is neither a finite upper bound nor a finite lower bound on the values that *des* assigns to propositions in the preference ranking, condition (6-5) forces c to be zero, and equations (6-6) and (6-7) assume the forms

$$DES\; X = des\; X$$

$$PROB\; X = prob\; X .$$

For if *des X* can assume arbitrarily large negative values, no positive value of c is small enough to ensure that the left-hand side of (6-5) will always be greater than the right; and if *des X* assumes arbitrarily large positive values, no negative value of c is close enough to zero to ensure that the left-hand side of (6-5) will always be greater than the right. *If des is unbounded both above and below, condition* (6-4) *implies that DES* = *des and PROB* = *prob.*

But if *des* is bounded above or below or both, c need not be zero. Here the natural question is, Given that *des* has a certain range of values, what values of c are possible consistently with condition (6-5)? (Each such possible value of c will determine a new pair of assignments *PROB, DES,* via equations [6-6] and [6-7].) An answer will now be given with the aid of condition (6-7).

Suppose that the preference ranking has neither a top nor a bottom, that there is no proposition that is ranked at least as high as every proposition in the preference ranking, and that there is no proposition that is ranked at least as low as every proposition in the ranking. Then there will be some number s which is either finite or is ∞ or $-\infty$ and which has the characteristic of being greater than *des X,* no matter what proposition in the preference ranking X may be, while no smaller number has that characteristic. The number s will be called the *supremum,* or *least upper bound,* of the values of *des.* Similarly, there will be a number i, the *infimum,* or *greatest lower bound,* of the values of *des.* The infimum has the characteristic of being smaller than all the values of *des,* while no larger number has that characteristic. Now there are four possibilities.

(a) $s = \infty$ and $i = -\infty$: *des* is unbounded above and below.
(b) $s = \infty$ and $i \neq -\infty$: *des* is unbounded above, bounded below.
(c) $s \neq \infty$ and $i = -\infty$: *des* is bounded above, unbounded below.
(d) $s \neq \infty$ and $i \neq -\infty$: *des* is bounded above and blow.

If X is a good proposition, *des X* will be positive, and (6-5) can be written as

$$c > -\frac{1}{des\ X}.$$

Since this inequality must hold no matter how high the proposition X is in the preference ranking, it must be the case that

$$c \geqslant -\frac{1}{s}.$$

Similarly, if X is bad, *des X* will be negative, and (6-5) can be written as

$$c < -\frac{1}{des\ X}.$$

Since this inequality must hold no matter how low the proposition X is in the preference ranking, it must be the case that

$$c \leqslant -\frac{1}{i}.$$

Combining the two conditions, we now have

$$-\frac{1}{s} \leqslant c \leqslant -\frac{1}{i}. \tag{6-10}$$

Since *des* $G = 1$, s must be greater than 1, and accordingly, we have

(6-11) $$c \geqslant -1$$

for every preference ranking that conforms with the general assumption (6-4).

6.4 Bounds on *c*

The values attainable by *des X* may be represented by the points on a vertical line or line segment, as in figures 6.1 (a), (b), (c), or (d). In figure 6.1 (a) we have the case in which *des X* can assume all values between $-\frac{1}{3}$ and $+2$: there, $s = 2$ and $i = -\frac{1}{3}$. In fugure 6.1 (b), *des X* can assume all values less than $+2$, and there is no lower bound on the finite negative values that are attainable: $s = 2$, $i = -\infty$. In figure 6.1 (c), all

Fig. 6.1

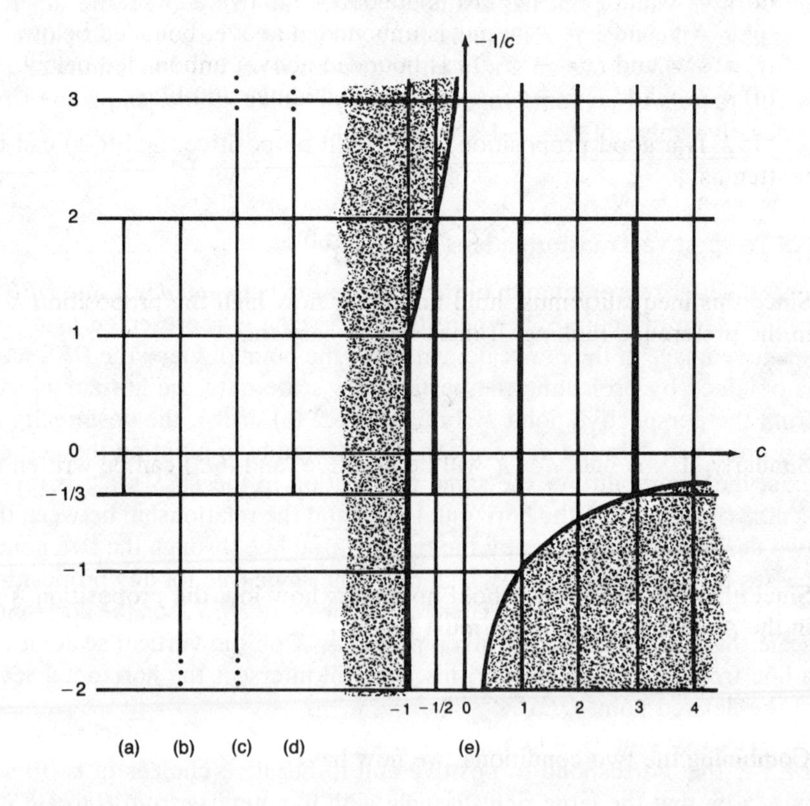

finite positive and negative values are attainable: $s = \infty$, $i = -\infty$. Finally, in figure 6.1 (d), all finite values greater than $-1/3$ are attainable: $s = \infty$, $i = -1/3$. What freedom is there in choosing c, given one of these ranges? The answer is given by condition (6-10). It can also be found graphically by sliding the vertical line segment that represents the range in question along, parallel to the vertical axis of figure 6.1 (e) until it is in a position where no part of it touches any portion of the shaded area. The line segment in figure 6.1 (a) will fit at any position from $c = -1/2$ to $c = +3$: three such positions for it are drawn in figure 6.1 (e), corresponding to $c = -1/2$, $c = +1$ and $c = +3$. The half line shown in figure 6.1 (b) will fit at any position from $c = -1/2$ to $c = 0$. The full line in figure 6:1 (c) will fit at only one position: $c = 0$. Finally the half line shown in figure 6.1 (d) will fit anywhere from $c = 0$ to $c = +3$. In general, the line segment representing the range will fit anywhere from $c = -1/s$ to $c = -1/i$.

Apparently, the choice $c = 0$ is always possible. If *des X* can attain all positive values, but has a finite infimum, there will be some range of possible values for c, extending from $c = 0$ to a positive value of $c = -1/i$. Also, if *des X* can attain all negative values but has a finite supremum, there will be some range of possible values for c, extending from a negative value of $c = -1/s$ to the value $c = 0$.

6.5 Perspective Transformations of Desirability

A better representation of the relationship between *des X* and *DES X* can be seen in figure 6.2, where the *DES* scale is represented as a perspective image of the *des* scale, with *P* as the point of view: the *DES* scale is obtained by projecting the vertical *des* scale onto the horizontal axis from the perspective point *P*. In figures 6.2 (a) or (b), the desirability of a proposition *X* in the *des* scale is represented by a point on the vertical axis; the desirability of the same proposition in the *DES* scale is represented by a point on the horizontal axis; and the relationship between the two desirabilities is given by the fact that the line through the two points passes through the point *P*, which plays the same role for any proposition *X* in the preference ranking. Then to find the point *DES X* on the horizontal scale that corresponds to a given point *des X* on the vertical scale, draw a line from *P* through *des X:* this line will intersect the horizontal scale at the desired point *DES X*.

The two graphs in figure 6.2 represent transformations of *des X* into *DES X* that correspond to positive and to negative choices of c. To see this, note that the large right triangle with hypotenuse from *P* to *DES X* in figure 6.2 (a) is similar to the small right triangle with hypotenuse from

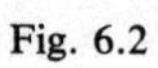
Fig. 6.2

(a)

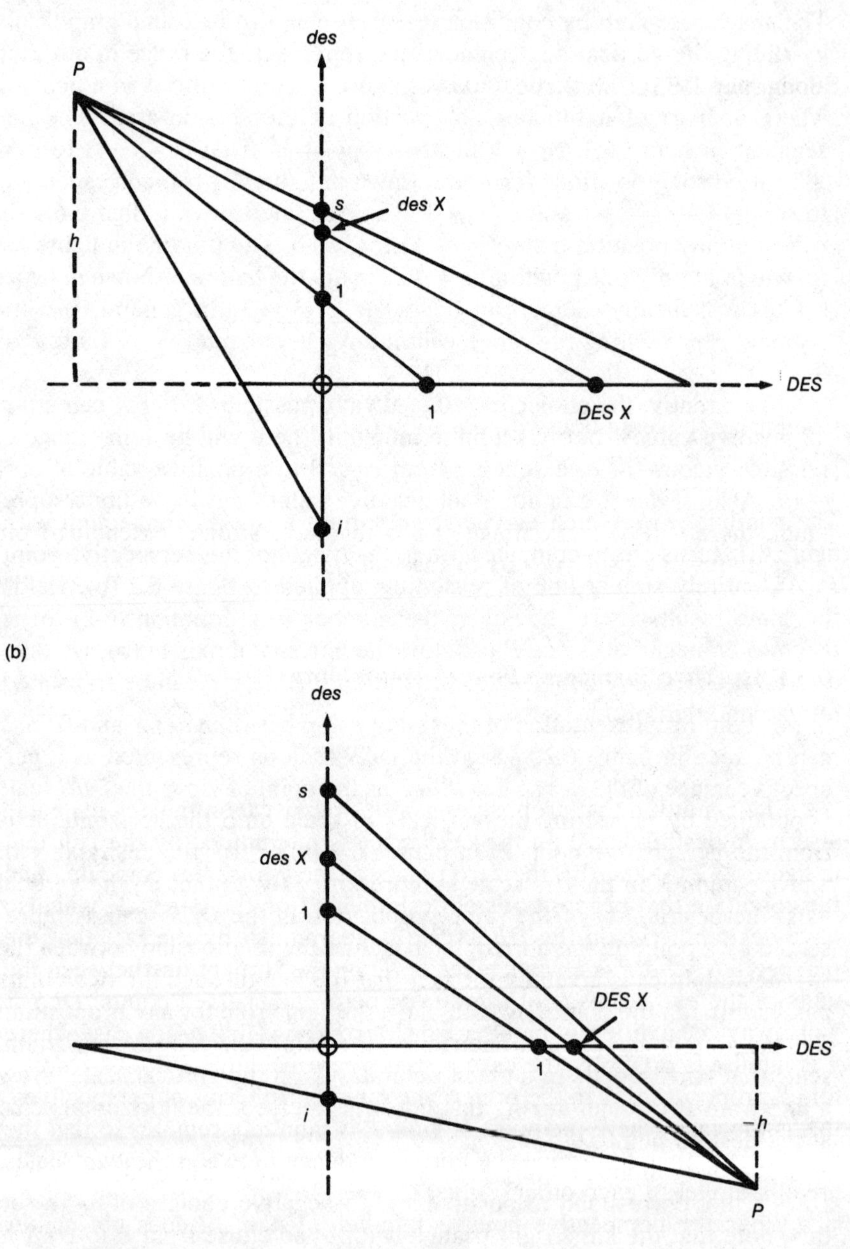
des
P
s
des X
h
1
DES
1
DES X
i
(b)
des
s
des X
1
DES X
DES
1
i
−h
P

des X to *DES X*. Therefore the base of the small triangle is to its height as the base of the large triangle is to its height:

$$\frac{DES\ X}{des\ X} = \frac{DES\ X + h - 1}{h}$$

Solving for *DES X* we have

$$DES\ X = \frac{(h - 1)(des\ X)}{-(des\ X) + h}$$

or, dividing by *h*,

$$DES\ X = \frac{\left(1 - \frac{1}{h}\right)(des\ X)}{-\frac{1}{h}(des\ X) + 1}$$

Comparing this equation with equation (6-6), we see that

(6-12) $$h = -1/c\ .$$

The number $-1/c$ which played so important a role in connection with figure 6.1, turns out to coincide with *h*, the height of the perspective point *P*. An entirely similar line of reasoning, applied to figure 6.2 (b), yields the same result where, however, the number *h* in equation (6-2) (b) is taken to be negative: since *P* is below the horizontal axis in (b), we take it to have a negative height. Now condition (6-10) on *c* can be translated into a condition on *h*:

(6-13) $$\text{either } h \leqslant i \text{ or } h \geqslant s\ .$$

The condition that propositions of desirability 0 on the *des* scale will also have desirability 0 on the *DES* scale is guaranteed by the fact that in figures 6.2 (a) and (b), the zero points of the two scales coincide. And the condition that propositions of desirability 1 on the *des* scale will also have desirability 1 on the *DES* scale is guaranteed by the fact that the perspective point *P* is always taken to lie on the straight line between the units of the two scales. But there need be no other fixed points: *DES X* will always be a different number from *des X* if *h* is finite. The case where *h* is infinite is precisely the case $c = 0$ where we have $DES\ X = des\ X$ for all propositions *X* in the preference ranking. In terms of perspectives, this is the case where the point of view *P* is infinitely remote, so that the rays from *P* that determine the correspondence between the two scales are all parallel to each other.

When the perspective point is infinitely distant, it does not matter whether *h* is taken to be $+\infty$ or $-\infty$: the rays will be parallel in either case. However, where h is finite there is an important difference between the cases where P is above the horizontal axis and those where P is below it. This is precisely the difference between positive and negative values of h: between negative and positive values of c, which has the opposite sign from h.

Thus, in figure 6.2 (a), h is positive and therefore c is negative; in figure 6.2 (b), c is positive. In general, when c is negative, *DES X* is greater than *des X* if X is prefered to G, and *DES X* is (algebraically) less than *des X* if X is liked less than G. Figure 6.2 (b) shows that for negative c the situation is reversed: in passing from *des X* to *DES X*, the scale of desirabilities above G is compressed, while the scale below G is expanded.

As condition (6-13) indicates, there is a lower limit on the allowable heights of P in figure 6.2 (a): h may be no smaller than s. With $h = s$, the supremum itself is projected to infinity on the *DES* scale. Similarly, with $h = i$ in figure 6.2 (b), the infimum i is projected to negative inifinity on the *DES* scale. Then a *des* scale on which there is only a finite range of attainable desirabilities can be transformed by equation (6-6) into a scale in which there is no upper bound on the attainable values but in which there is a lower bound; or it can be transformed by that equation into a scale in which there is no lower bound on the attainable values but in which there is an upper bound. It is impossible, however, to transform a finite scale into a doubly infinite one: after the transformation there will be an upper bound or a lower bound on the attainable desirabilities—perhaps both, and at least one.

6.6 Probability Quantization

The surprising feature of transformation (6-1) is that *PROB X* may depend on *des X* as well as on *prob X*: our system differs from Ramsey's in that for us, the preference ranking need not uniquely determine the probability assignment. (See chapter 10 for the reason why.) In analyzing this situation it will again be convenient to assume that the *des* and *DES* scales agree on the values they assign to two differently ranked propositions as in (6-4). In this way we use up the latitude in the choice of the desirability scale that the present system shares with Ramsey's and thus focus attention on the peculiar features of the present system. Then the probability and desirability transformations are given by equations (6-7) and (6-6), where the parameter c must satisfy condition (6-5).

How much is the probability of a proposition X changed by a transformation of form (6-7)? The answer depends on the parameter c, and the allowable values of c are given by condition (6-10). With c at its maximum value of $-1/i$, the new probability of X given by (6-7) is

$$(prob\ X)\left(1 - \frac{des\ X}{i}\right)$$

which is maximum or minimum depending on whether X is good or bad. And with c at its minimum value of $-\ 1/s$, the new probability of X given by (6-7) is

$$(prob\ X)\left(1 - \frac{desX}{s}\right)$$

which is minimum or maximum depending on whether X is good or bad. The difference between these extremes is

$$prob\ X\ des\ X\left(\frac{1}{s} - \frac{1}{i}\right). \tag{6-13}$$

The probability variation (6-13) will be 0 if X has probability 0 or if X is indifferent. But how great might (6-13) be? The answer depends on the values attained by the quantity *prob X des X* for various values of X. To study the situation it is convenient to define

(6-14)
$$\text{(a) } int\ X = prob\ X\ des\ X$$
$$\text{(b) } INT\ X = PROB\ X\ DES\ X$$

where "*int*" is short for *integral*. (The reason for choosing this term will appear shortly.) Applying definitions (6-14) to the general transformations (6-1) and (6-2), we have the transformations

(6-15)
$$\text{(a) } INT\ X = a\ int\ X + b\ prob\ X$$
$$\text{(b) } PROB\ X = c\ int\ X + d\ prob\ X$$

for *prob* and *int* in the general case; and under the special assumption (6-4), we have

(6-16)
$$\text{(a) } INT\ X = (c + 1)\ int\ X$$
$$\text{(b) } PROB\ X = c\ int\ X + prob\ X$$

By the desirability axiom (5-2), *int* is additive (as is *INT*):

(6-17)
$$\text{(a) If } XY = F, \text{ then } int(X \vee Y) = int\ X + int\ Y$$
$$\text{(b) If } XY = F, \text{ then } INT(X \vee Y) = INT\ X + INT\ Y$$

The absolute value of the variation (6-13) must be less than 1, for a proposition has probability 0 on both the *prob* and *PROB* scales if it has probability 0 on either. Then by definition (6-14) (a) we have

$$int\ X < \frac{1}{(1/s) - (1/i)} \tag{6-18}$$

Now since the denominator on the right-hand side of (6-18) cannot be 0, *INT* must be bounded above and below, whether or not *DES* is.

Furthermore, the infimum of *int* (say, i'), must be equal to the supremum s', of *int*, with its sign changed: $i' = -s'$. To see this, note that by (6-14) (a), we have *int* $T = 0$, so that, setting $Y = \overline{X}$ in (6-17) (a), we have

$$\text{(6-19)} \qquad int\ \overline{X} = -int\ X$$

for every proposition X in the preference ranking. Now suppose that the sequence of numbers

$$int\ X_1,\ int\ X_2,\ \ldots$$

increases toward s' as a limit. Then the sequence

$$int\ \overline{X}_1,\ int\ \overline{X}_2,\ \ldots$$

must converge toward $-s'$ as a limit. If $-s'$ were not i' (and were therefore greater than i') there would be some decreasing sequence

$$int\ Y_1,\ int\ Y_2,\ \ldots$$

that converged toward i', and therefore there would be an n such that *int* Y_n would be less than $-s'$. By (6-19) *int* $\overline{Y}_n$ would then have to be greater than s; but this is impossible, since s' is the supremum of *int*. Then $-s'$ must equal i' as claimed.

Now the pairs of values (*prob* x, *int* X) that are actually attained by propositions X in the preference ranking can be represented by the points in or on a closed curve of the sort shown in figure 6.3. The case depicted in figure 6.3 is one in which s', the supremum of *int*, is 1; s, the supremum of *des*, is 3; and i, the infimum of *des*, is -6. The radial symmetry of the curve about the point (1/2, 0), is accounted for by (6-19) together with the fact that *prob* $\overline{X} = 1 -$ *prob* X. Each ray from the origin in figure 6.3 contains all points representing propositions of a fixed desirability: each such ray is an *indifference curve*. The effect of the transformations (6-16) can be indicated by capitalizing the labels "*prob*," "*des*," and *int* in figure 6.3 and suitably distorting the closed curve. The radial symmetry about the point (1/2, 0) will be retained, and points on the same ray from the origin in figure 6.3 will again be on a single ray from the origin after the distortion, although with the exception of the rays corresponding to *des* $= 0$ and *des* $= 1$, these rays will go through various degrees of rotation in the course of the distortion.

Now the probability variation (6-13) will approach its maximum absolute value for propositions X for which *int* X approaches its maximum or minimum. Thus the supremum of the absolute probability variation will be

Fig. 6.3

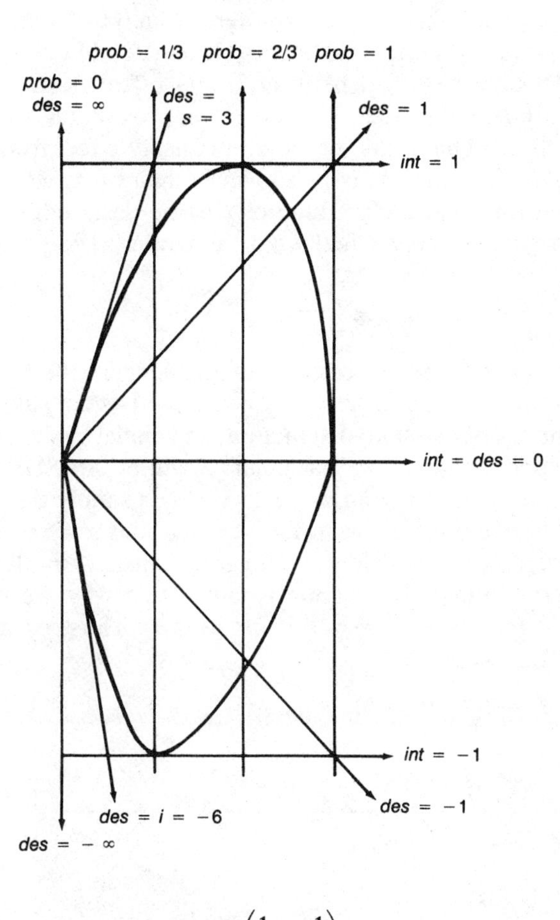

(6-20) $$s'\left(\frac{1}{s} - \frac{1}{i}\right)$$

where s' is the supremum of *int*, s is the supremum of *des*, and i is the infimum of *des*. In the case depicted in figure 6.3, the maximum absolute probability variation is 1/6 and is attained by propositions of (a) desirability 3/2 and probability 2/3, and (b) desirability −3 and probability 1/3.

Example 1

A marksman fires repeatedly at a distant circular target one foot in radius until he hits it, at a distance x feet from the center. The agent will pay the marksman \$1.50 for each inch by which x is less than four, and the marksman will pay the agent \$.375 for each inch by which x exceeds four. Assume that the agent identifies desirabilities with numbers of dollars gained, counting losses as negative gains; and assume that in the agent's opinion, x is equally likely to be anywhere from 0 to 1 foot.

To be more precise, if $0 \leqslant a < b \leqslant 1$, let $A(a, b)$ be the proposition that the bullet hits the target between a and b feet from the center, so that for small positive e, $A(x - e, x + e)$ approximates the proposition that the (idealized) bullet hits the target precisely x feet from the center. Then the desirability picture is as shown in figure 6.4, where the height y of the graph for a particular value of x is the desirability of the bullet hitting the target precisely x feet from the center. The probabilities are described by the statement that

$$prob\ A(a,b) = b - a$$

if $0 \leqslant a < b \leqslant 1$. Let the preference ranking concern the propositions of form $A(a, b)$ with a and b as above, together with the results of applying the operations of conjunction, disjunction, and denial to such propositions any finite number of times. In particular, T will be $A(0, 1)$.

Now the integral of y from $x = a$ to $x = b$ is simply the total amount of shaded area enclosed between the vertical lines $x = a$ and $x = b$ in figure 6.4, counting areas below the horizontal axis ($y = 0$) as negative. When the term "integral" is understood in this way we find that the integral of y from $x = a$ to $x = b$ is $int\ A(a, b)$. Thus, by the additivity of int, (6-17) (a), we have

$$int\ T = int\ A(0, 1/3) + int\ A(1/3, 1) = -1 + 1 = 0$$

Fig. 6.4

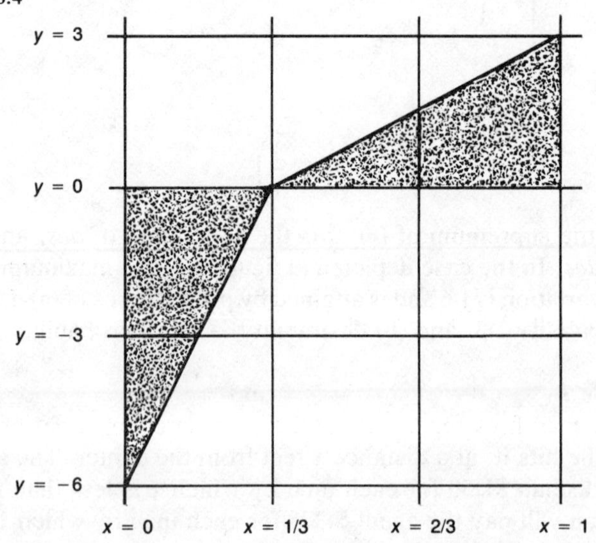

since the area of a triangle is half the product of its base and height.

In terms of figure 6.4, *des* $A(a, b)$ is the average value of y between $x = a$ and $x = b$. Thus, *des* $A(0, 1/3)$ is -3 and *des* $A(1/3, 1)$ is 1.5. This agrees with the general stipulation that *des* X = *int* X/*prob* X.

The attainable pairs of values (*int* X, *prob* X), with X in the probability field described above, are as shown in figure 6.3. In particular, *int* X is maximum for $X = A(1/3, 1)$ and minimum for $X = A(0, 1/3)$.

6.7 Problems

1

Change example 1 by assuming that the marksman keeps firing until the bullet hits the target at a distance x that is not precisely four inches from the center; that the agent then pays the marksman six dollars if x is less than 1/3 of a foot; and that the marksman pays the agent three dollars if x is more than 1/3 of a foot. (Keep the same assumptions about probabilities.) Draw graphs corresponding to figures 6.4 and 6.3 under the new assumptions.

2

In the preference ranking of example 1, there is no proposition X for which *des* $X = 3 = s$ or for which *des* $X = -6 = i$. Is this still the case when the example is changed as in problem 1? Explain.

3

Change example 1 by assuming that the agent pays the marksman 4/6 dollars for each inch by which x is less than six, and that the marksman pays the agent 4/6 dollars for each inch by which x exceeds six. Draw graphs corresponding to figures 6.4 and 6.3. Transform *prob, des,* and *int* via (6-6) and (6-15) with $c = -1/s$, and suitably redraw the graphs corresponding to figures 6.4 and 6.3.

4

Explain why each of the three conditions (6-3) must be satisfied for the equivalence theorem to hold.

5

In discussing transformations (6-6) and (6-7) we assumed that the preference ranking had no top and no bottom because we wanted all desirabilities to be finite. Explain how it comes about that if there is a top-ranking proposition or a bottom-ranking proposition, its desirability can be changed to ∞ or to $-\infty$ by transformations of the sort described in the equivalence theorem.

6

How (if at all) would the function *des* in example 1 be affected if the graph in figure 6.4 were altered by setting $y = -6$ when $x = 2/3$? (For values of x other than 2/3, the values of y remain those shown in figure 6.4.)

6.8 Acknowledgment

"The equivalence theorem" is my name for an easy fragment of a theorem of Ethan Bolker's that he explained to me in 1964. The rest of the fragment is a uniqueness theorem, of which the one proved in chapter 8 is a variant. I also heard of the equivalence theorem from Kurt Gödel, who had noticed it independently, after Bolker had proved it and the uniqueness theorem. The proof given here is my account of the way Bolker explained it to me in 1964. Bolker's published proofs are far more compact and elegant, but are addressed to people of greater mathematical attainments than I have or suppose my readers to have. For references, see section 8.8.

7

From Preference to Probability

To what extent does an agent's preference ranking determine his probability assignment? In section 7.1 we note four conditions on preference rankings under which (as we prove in the rest of the chapter) the ratio of probabilities of any two propositions that are ranked together is uniquely determined. This result, (7-5), will play a central role in the uniqueness proof of chapter 8. We begin by considering examples where preferences are more or less revealing of probabilities.

Example 1: Total indifference

If all propositions in the collection are ranked together, then all of them have the same numerical desirability, and nothing can be concluded about their probabilities beyond such general truths as are embodied in the probability axioms (5-1).

Example 2: Zero probability

It is rather easy to identify good or bad propositions that have probability 0. If the relevant portion of the preference ranking is (a) or (b) below, the proposition A must have probability 0; and conversely, if A is good and has probability 0, the relevant portion of the preference ranking must be as shown in (a), while if A is bad and has probability 0, the relevant portion of the preference ranking must be as in (b).

(a)	(b)
A	$\bar{A}, T$
$\bar{A}, T$	A

To see that this is the case, set $X = A$ and $des\ T = 0$ in equation (5-3):

$$0 = prob\ A\ des\ A + prob\ \bar{A}\ des\ \bar{A}\ .$$

In (a) and in (b) we have $des\ \overline{A} = des\ T = 0$ so that the equation becomes

$$0 = prob\ A\ des\ A\ .$$

Since A is not ranked with T, $des\ A$ is not 0, and therefore we can divide both sides by it to get

$$0 = prob\ A.$$

Example 3: Probability comparisons

Consider the following three preference rankings, in each of which A and B are ranked together, but in which the relative positions of their denials vary.

(a)	(b)	(c)
A, B	A, B	A, B
T	T	T
$\overline{A}, \overline{B}$	$\overline{A}$	$\overline{B}$
	$\overline{B}$	$\overline{A}$

in these three rankings the relationships between the probabilities of A and of B must be

(a) $prob\ A = prob\ B$ (b) $prob\ A < prob\ B$ (c) $prob\ A > prob\ B$

The easiest way to see that this is the case is to consider the meaning of equation (5.4). The numerator,

$$des\ T - des\ \overline{X}\ ,$$

is the desirability interval from $\overline{X}$ to T, and the denominator,

$$des\ X - des\ \overline{X}\ ,$$

is the desirability interval from $\overline{X}$ to X. Then the equation can be read as follows:

$$prob\ X = \frac{\text{desirability interval from } \overline{X} \text{ to } T}{\text{desirability interval from } \overline{X} \text{ to } X}\ .$$

In preference ranking (a) above, this ratio will be the same whether we set $X = A$ or $X = B$, since $\overline{A}$ is ranked with $\overline{B}$ and A is ranked with B. In (b), the ratio

$$\frac{\text{desirability interval from } \overline{A} \text{ to } T}{\text{desirability interval from } \overline{A} \text{ to } A}$$

is smaller than the ratio

$$\frac{\text{desirability interval from } \overline{B} \text{ to } T}{\text{desirability interval from } \overline{B} \text{ to } B}$$

and, consequently, we have *prob* $A <$ *prob* B, as claimed. In (c), the second ratio is the smaller, so that we have *prob* $A >$ *prob* B. By similar reasoning, it can be shown that for the preference rankings

(d)	(e)
$\overline{B}$	$\overline{A}$
$\overline{A}$	$\overline{B}$
T	T
A, B	A, B

the probability relationships must be

(d) *prob* $A <$ *prob* B (e) *prob* $A >$ *prob* B .

For a concrete example of the ranking (c), suppose that the Bayesian agent has applied for two jobs, would be equally pleased to get either, but thinks he is much more likely to (A) get job 1 than to (B) get job 2. It follows that he would be more displeased to learn ($\overline{A}$) that he has failed to get job 1 than to learn ($\overline{B}$) that he has failed to get job 2: to the extent that he expected failure, he had written it off in advance. His high degree of belief in $\overline{B}$ has already colored his view of the necessary proposition, $T = B \vee \overline{B}$, so that $\overline{B}$ does not seem as bad (= as far below T) as $\overline{A}$.

The situations discussed in examples 2 and 3 can be summarized by the following statement.

(7-1) Of two propositions which are ranked together but not with T, the more probable is the one whose denial is further from T in the preference ranking.

This implies that if the necessary proposition is assigned desirability 0, and if

$$des\ X = des\ Y \neq 0 ,$$

then the more probable proposition is the one whose denial has the greater numerical desirability, ignoring signs. By the same reasoning, we can compare the probabilities of propositions whose denials are ranked together but not with T. The more probable of two such propositions is the one which is closer to T in the preference ranking: the one with the smaller numerical desirability. This allows us to recognize A as more probable than B in the first two of the following rankings, and to recognize A as less probable than B in the second two.

$\overline{A}, \overline{B}$	B	$\overline{A}, \overline{B}$	A
T	A	T	B
A	T	B	T
B	$\overline{A}, \overline{B}$	A	$\overline{A}, \overline{B}$

7.1 The Existence, Closure, G, and Splitting Conditions

From these examples it should be evident that certain conditions must be met if we are to be able to deduce features of the agent's probability assignment from his preference ranking. Obviously, one indispensable condition is that there be a desirability function that represents the agent's preferences and is associated via the desirability axiom (5-2) with some probability function.

EXISTENCE CONDITION. There is a pair of assignments *prob, des* that satisfy the probability axioms (5-1) and the desirability axiom (5-2), with *des* mirroring the preference ranking in the sense that for any two propositions *A, B* in the ranking, *des A* is greater than, equal to, or less than *des B* depending on whether *A* is ranked above, with, or below *B*.

In formulating the second condition it will be helpful to use the term *probability field,* defined as follows:

A probability field is a nonempty collection of propositions that contains the denial of any proposition that it contains, and contains the conjunction and the disjunction of any pair of propositions that it contains.

Thus, if A is in a probability field, so is $\overline{A}$, and if A and B are in a probability field, so are AB and $A \vee B$. Then if C is also in the field, so is the three-termed conjunction *ABC;* and so on. Probability fields are so called because it is to such collections of propositions that probabilities are most naturally and conveniently attributed. Furthermore, it is to just such collections that desirabilities are most conveniently attributed, except that it is pointless to attribute any desirability to the impossible proposition—even, desirability 0.

CLOSURE CONDITION. The propositions in the agent's preference ranking form a probability field from which the impossible proposition has been removed.

As its name suggests, this condition requires the collection of propositions in the agent's preference ranking to be *closed* under certain operations.

In particular, the condition posits closure under the operations of denial, conjunction, and disjunction—except where these operations yield the impossible proposition. Thus, closure under disjunction ensures that $A \vee B$ is in the ranking whenever A and B both are, for disjunction cannot yield an impossibility if both disjuncts are possible; but closure under conjunction must be restricted, e.g., because $A\overline{A} = F$, and similarly, closure under denial must be restricted because $\overline{T} = F$.

The closure condition would not be violated if the preference ranking contained just one proposition, T. We shall want to exclude this trivial case. We shall also want to exclude the case where, although the necessary proposition is not alone in the ranking, all propositions are indifferent in the sense of being ranked with T. And we shall also want to exclude the almost equally trivial case in which the only nonindifferent propositions have probability 0. This is accomplished by the third condition.

G CONDITION. In the preference ranking there is a good proposition, G, of which the denial is bad.

Since a good proposition is one that is preferred to T, and a bad proposition is one to which T is preferred, the G condition asserts that there is in the preference ranking a proposition G for which the relevant part of the ranking is as follows.

$$G$$

$$T$$

$$\overline{G}$$

As example 2 suggests, neither G nor $\overline{G}$ can have probability 0. Indeed, if *prob* G were 0, equation (5-3) would yield

$$des\ T = (0)(des\ G) + (1)(des\ \overline{G}) = des\ \overline{G}$$

so that $\overline{G}$ would be ranked with T. A similar argument shows that G is ranked with T if *prob* $\overline{G}$ is 0. Consequently, the G condition implies that the probability of G is neither 0 nor 1.

Our fourth and final condition says in effect that any good or bad proposition A of positive probability can be split into two equiprobable parts, A_1, A_2, that are ranked with each other (and, therefore, with A itself).

SPLITTING CONDITION. If A is in the ranking, and neither A nor its denial is ranked with T, then there will be propositions A_1, A_2 in the ranking that satisfy the following conditions.

(1) $A_1 A_2 = F$;
(2) $A_1 \vee A_2 = A$;

(3) A_1 and A_2 are ranked together;
(4) $\overline{A}_1$ and $\overline{A}_2$ are ranked together.

By example 2, the condition that neither A nor $\overline{A}$ be ranked with T ensures that the probability of A is not zero; and conversely, if A is not ranked with T and does not have probability zero, then $\overline{A}$ will not be ranked with T. By example 3, conditions (3) and (4) imply that A_1 and A_2 have the same probability, and by (1) and (2), this probability is half the probability of A. Also, by (1), (2), (3), and the fact that A does not have zero probability, we have $des\ A_1 = des\ A_2$ and

$$des\ A = \frac{prob\ A_1\ des\ A_1 + prob\ A_2\ des\ A_1}{prob\ A_1 + prob\ A_2} = des\ A_1$$

so that A, A_1, and A_2 are all ranked together.

The splitting condition is plausible if the agent believes that there are various well-balanced coins or standard decks of cards in the world. Thus, the proposition (A) that

tomorrow will be a fine day

can be expressed as the disjunction of the proposition (A_1) that

tomorrow will be a fine day day and the next card dealt in Las Vegas will be red

and the proposition (A_2) that

tomorrow will be a fine day and the next card dealt in Las Vegas will be black.

It is entirely plausible that the splitting condition is met for this choice of A, A_1, A_2; and it is plausible that the same sort of device can be applied to any good or bad proposition A to find a pair A_1, A_2 which, with A, satisfies the splitting condition.

7.2 Determining Ratios of Probabilities

The splitting condition implies that any good or bad proposition in the preference ranking which has positive probability can be split into two equiprobable parts, each of which can then be split into two equiprobable parts, etc. Then such a proposition can be split into 2 or 4 or 8 or, in general, 2^n equiprobable parts, each of which is liked exactly as well as the original proposition. This means that we can compare the probabilities of any two good propositions that are ranked together, and we can compare the probabilities of any two bad propositions that are ranked together. We can make these comparisons in the sense that we can determine the ratio

$$\frac{prob\ A}{prob\ B}$$

as accurately as we please, provided only that A and B have the same position in the agent's preference ranking, and that this be a different position from that occupied by the necessary proposition T.

To make such comparisons we must be able to recognize which of the cases

$$prob\ X > prob\ Y \quad prob\ X = prob\ Y \quad prob\ X < prob\ Y$$

holds, whenever X and Y are ranked with each other but not with T. This can be done with the aid of (7-1). Now, to determine the ratio

$$\frac{prob\ A}{prob\ B}$$

to a certain degree of accuracy, determine whether A and B have the same probability. If they do, the ratio is 1, and the problem is solved. If they do not, find which is the more probable. Suppose for definiteness that A and B are both good and that A is less probable than B but does not have probability 0. This situation will be revealed by the fact that the relevant portion of the agent's preference ranking appears as follows.

$$\begin{array}{c} A, B \\ T \\ \overline{A} \\ \overline{B} \end{array}$$

We know that the ratio is between 0 and 1:

$$0 < \frac{prob\ A}{prob\ B} < 1\,.$$

The ratio is less than 1 because A is less probable than B, greater than 0 because A has positive probability, and determinate because B also has positive probability. To locate the ratio more precisely within the interval, split B, then split the two pieces of B, and then split those four pieces of B, and so on. For definiteness, suppose the splitting procedure has been applied ten times, yielding after the successive applications

2, 4, 8, 16, 32, 64, 128, 256, 512, 1024

equiprobable incompatible propositions whose disjunctions are B and which are all ranked with B. Then after the tenth splitting, we have 1,024 propositions,

$$B_1, \ldots, B_{1,024}$$

each of which has probability

$$\frac{prob\ B}{1{,}024}\ .$$

The disjunction of the first n of these propositions will have probability

$$prob\ (B_1 \vee B_2 \vee \ldots \vee B_n) = \frac{n}{1{,}024}\ prob\ B\ ,$$

where n can be any number you like, from 1 to 1,024. Suppose that in fact, the probability of A is between 18/1,024 and 19/1,024 times the probability of B. This fact will reveal itself in the preference ranking: the proposition $\overline{A}$ will be ranked between the proposition

$$\overline{B_1 \vee B_2 \vee \ldots \vee B_{18}}$$

and the proposition

$$\overline{B_1 \vee B_2 \vee \ldots \vee B_{18} \vee B_{19}}$$

in the agent's preference ranking. We would then know that the ratio lies within a certain interval of length 1/1,024:

$$\frac{18}{1{,}024} < \frac{prob\ A}{prob\ B} < \frac{19}{1{,}024}\ .$$

In other words, we should have determined the ratio to within an accuracy of better than one part in a thousand. Had twice, or four times this accuracy been needed, we could have obtained it by applying the splitting procedure once or twice more.

Then the ratio of the probabilities of two propositions which are ranked with each other but not with T is completely determined by the agent's preference ranking of propositions: witness, the fact that we can determine the ratio as accurately as we please, by going through an appropriately large but finite number of steps of a simple procedure.

Example 4

Suppose that the splitting procedure has been applied to B three times, yielding eight equiprobable slices of B. This might be accomplished by reference to three tosses of a coin. If H_n is the proposition that the coin lands head up on the nth toss, the eight slices of B might be as follows:

$$B_1 = BH_1H_2H_3$$

$$B_2 = BH_1H_2\overline{H}_3$$

$$B_3 = BH_1\overline{H}_2H_3$$

$$B_4 = BH_1\overline{H}_2\overline{H}_3$$
$$B_5 = B\overline{H}_1H_2H_3$$
$$B_6 = B\overline{H}_1H_2\overline{H}_3$$
$$B_7 = B\overline{H}_1\overline{H}_2H_3$$
$$B_8 = B\overline{H}_1\overline{H}_2\overline{H}_3$$

The first application of the splitting procedure would have yielded two slices of B: BH_1 and $B\overline{H}_1$, i.e., $B_1 \vee B_2 \vee B_3 \vee B_4$ and $B_5 \vee B_6 \vee B_7 \vee B_8$. The second application would have split the first of these slices into BH_1H_2 (which is $B_1 \vee B_2$) and $BH_1\overline{H}_2$ (which is $B_3 \vee B_4$) and, similarly, would have split the second of the slices into $B_5 \vee B_6$ and $B_7 \vee B_8$. The eight B's listed above would then have resulted from splitting each of these four slices, via H_3.

It goes without saying that splitting is not a physical operation: nothing is done really to B when you "split" it into eight "slices." Rather, the splitting procedure consists in finding eight propositions of the required sort in the agent's preference ranking. It may well be that B_1 through B_8, as defined above, are eight such propositions.

Now, regardless of which propositions are made to play the role of the eight slices of B, suppose that the preference ranking is as follows.

$$A, B, B_1, B_2, B_3, B_4, B_5, B_6, B_7, B_8$$
$$T$$
$$\overline{B_1}$$
$$\overline{B_1 \vee B_2}$$
$$\overline{B_1 \vee B_2 \vee B_3}$$
$$\overline{B_1 \vee B_2 \vee B_3 \vee B_4}$$
$$\overline{B_1 \vee B_2 \vee B_3 \vee B_4 \vee B_5}$$
$$\overline{A}$$
$$\overline{B_1 \vee B_2 \vee B_3 \vee B_4 \vee B_5 \vee B_6}$$
$$\overline{B_1 \vee B_2 \vee B_3 \vee B_4 \vee B_5 \vee B_6 \vee B_7}$$
$$\overline{B}$$

Then the ratio has been determined to within an interval of length 1/8,

$$.625 < \frac{prob\ A}{prob\ B} < .750\ ,$$

for the disjunction of five of the slices of B has probability 5/8 that of B, i.e., .625 *prob B*, and the disjunction of six slices has probability 6/8 that of B, i.e., .750 *prob B*.

7.3 A Probability Scale for Indifferent Propositions

Thc splitting procedure cannot be applied directly to propositions that are ranked with T, but it can be applied indirectly, to determine the numerical probabilities (not just ratios of probabilities) of such propositions.

Example 5

Suppose the splitting procedure has been applied 10 times to the good proposition G whose existence is assumed in the G condition, to yield 1,024 incompatible equiprobable propositions

$$G_1, G_2, \ldots, G_{1{,}024}$$

whose disjunction is G and which are ranked with G. Also, suppose the procedure has been applied 10 times, to $\overline{G}$, yielding 1,024 equiprobable incompatible propositions

$$H_1, H_2, \ldots, H_{1{,}024}$$

whose disjunction is $\overline{G}$ and which are all ranked with $\overline{G}$. Now for each number n from 1 to 1,024, define a proposition

$$I_n = G_n \vee H_n .$$

Each such proposition will be ranked with T; different such propositions will be logically incompatible; all such propositions will have the same probability; and the disjunction of the 1,024 of them is T. Therefore we have

$$prob\ I_n = 1/1{,}024$$

for each n, and, consequently,

$$prob\ (I_1 \vee I_2 \vee \ldots \vee I_n) = n/1{,}024$$

Thus, the probability of an indifferent proposition is uniquely determined if it can be expressed as the disjunction of a certain number of slices of G with an equal number of slices of $\overline{G}$.

Even where an indifferent proposition does not have the special form $I_1 \vee I_2 \vee \ldots \vee I_n$ discussed in example 5, its probability can be determined to any desired degree of accuracy by comparing it with propositions of that special form. To carry out such comparisons we need a method which, unlike (7-1), can be applied to propositions that are ranked with *T;* but to formulate such a method, we need a technique for recognizing propositions of probability zero when we see them in the preference ranking—a technique which, unlike that of example 2, can be applied to propositions that are ranked with *T*.

7.4 Nullity

A technique of the required sort is implicit in the following definition of *nullity*.

(7-2) A proposition *A* in the preference ranking is *null* if and only if there is a proposition *B* in the preference ranking for which we have

(a) $AB = F$
(b) $A \vee B$ is ranked with *B*, and
(c) *B* is not ranked with *A*.

Now we can show that

(7-3) If the existence condition, the closure condition, and the *G* condition are satisfied, the null propositions in the preference ranking are precisely those to which *prob* assigns the value 0.

The proof of (7-3) has two parts. First, we prove that if *prob A* is not zero then *A* is not null; second, we prove that if *prob A is* zero then *A is* null.

For the first part of the proof, suppose that $prob\ A \neq 0$. To show that *A* is not null, we must then show that if $AB = F$, the other two conditions for nullity cannot both hold. (Note that if $A = T$ then for no *B* in the ranking do we have $AB = F$.) In particular, we shall suppose that $AB = F$ and that *B* is ranked with $A \vee B$, so that conditions (a) and (b) in the definition both hold; and we shall deduce from these assumptions that *A* and *B* are ranked together, so that condition (c) in the definition of nullity fails. By the assumptions we have

$$des\ B = des\ (A \vee B)$$

and

$$des\ (A \vee B) = \frac{prob\ A\ des\ A + prob\ B\ des\ B}{prob\ A + prob\ B}$$

where the denominator is not 0 because by assumption, $prob\ A \neq 0$. Combining these two equations, we have

$$des\ B = \frac{prob\ A\ des\ A + prob\ B\ des\ B}{prob\ A + prob\ B}$$

or, cross-multiplying and then subtracting *prob B des B* from both sides,

$$prob\ A\ des\ B = prob\ A\ des\ A\ .$$

Since $prob\ A \neq 0$, we can divide both sides by *prob A* to get

$$des\ B = des\ A\ .$$

Then A and B are ranked together, and condition (7-2) (c) fails. This finishes the first part of the proof.

For the second part of the proof, we suppose that $prob\ A = 0$ and then deduce that A is null. This involves producing a proposition B such that $AB = F$ and such that the preference ranking is either (a) or (b) below.

(a)	(b)
A	$B, A \vee B$
$B, A \vee B$	A

In fact, one of the following two choices will always work:

$$B_1 = \overline{A}G \quad B_2 = \overline{A}\overline{G}$$

To see this, notice first that both of the conditions

$$AB_1 = F \quad AB_2 = F$$

hold, so that whether we choose $B = B_1$ or $B = B_2$, B will be incompatible with A. To show that the preference ranking will be (a) or (b) above for at least one of these choices of B, we show first that B_1 and B_2 cannot both be ranked with A. To show this, note that since $G = AG \vee \overline{A}G$ we have

$$des\ G = \frac{prob\ AG\ des\ AG + prob\ \overline{A}G\ des\ \overline{A}G}{prob\ G}\ .$$

Since $prob\ A = 0$, we have $prob\ AG = 0$ and $prob\ \overline{A}G = prob\ G$ so that the equation becomes

$$des\ G = \frac{0 + prob\ G\ des\ \overline{A}G}{prob\ G} = des\ \overline{A}G\ .$$

Then $\overline{A}G$ is ranked with G. An entirely parallel argument shows that

$$des\ \overline{G} = \frac{0 + prob\ \overline{G}\ des\ \overline{A}\overline{G}}{prob\ \overline{G}} = des\ \overline{A}\overline{G}\ .$$

Then $\overline{A}\overline{G}$ is ranked with $\overline{G}$. Now by (5-2) the ranking must be

$$G, \bar{A}G$$

$$T$$

$$\bar{G}, \bar{A}\bar{G}$$

so that B_1 (i.e., $\bar{A}G$) and B_2 (i.e., $\bar{A}\bar{G}$) are certainly not ranked together. Then if A is ranked with B_1, it will certainly not be ranked with B_2. It only remains to show by (5-2) that for either choice of B we can be sure that B is ranked with $A \vee B$:

$$\begin{aligned} des\ (A \vee B_1) &= \frac{prob\ A\ des\ A + prob\ B_1\ des\ B_1}{prob\ A + prob\ B_1} \\ &= \frac{0 + prob\ B_1\ des\ B_1}{0 + prob\ B_1} \\ &= des\ B_1\ . \end{aligned}$$

Then B_1 is ranked with $A \vee B_1$; and an entirely parallel argument shows that

$$des\ (A \vee B_2) = des\ B_2\ .$$

This completes part two of the proof of (7-3).

7.5 A General Technique

Now we can describe a technique for comparing the probabilities of propositions A, B that are ranked together, whether they are ranked with T or not. The technique involves the use of a "test proposition," C.

(7-4) Suppose that the existence condition, the closure condition, and the G condition are all satisfied; that A and B are ranked together but not with a third proposition C; that $AC = BC = F$; and that C is not null. Then:

(a) $prob\ A = prob\ B$ if $A \vee C$ and $B \vee C$ are ranked together;
(b) $prob\ A < prob\ B$ if $A \vee C$ is ranked closer than $B \vee C$ to C;
(c) $prob\ A > prob\ B$ if $B \vee C$ is ranked closer than $A \vee C$ to C.

Intuitively, we may think of the rankings of $A \vee C$ and $B \vee C$ as determined by a pull that C exerts on A and on B against gravity, with probability playing the role of mass: the "lighter" (= less probable) of A, B is pulled closer to C than the "heavier."

To prove (7-4), let $des\ A = des\ B = x$, $des\ C = y$, $prob\ A = p$, $prob\ B = q$, $prob\ C = r$, and suppose that the hypotheses of (7-4) are satisfied, so that we have

$$x \neq y \qquad r \neq 0$$

By the desirability axiom (5-2), we have

$$des\ (A \vee C) = \frac{px + ry}{p + r}$$

$$des\ (B \vee C) = \frac{qx + ry}{q + r}$$

so that *des* $(A \vee C)$ and *des* $(B \vee C)$ are weighted averages of the desirabilities x, y, of A and of C,

$$des\ (A \vee C) = w_1x + w_2y$$

$$des\ (B \vee C) = w_3x + w_4y$$

where the ratios

$$w_1 : w_2 \qquad w_3 : w_4$$

of the weights in the two averages are, respectively,

$$p : r \qquad q : r$$

Now since $x \neq y$, the question of which (if either) of the numbers *des* $(A \vee C)$, *des* $(B \vee C)$ is the closer to y is determined by the ratios of the weights: the first number will be (a) equal to the second, or (b) closer than the second to y, or (c) further than the second from y accordingly as the ratio $w_1 : w_2$ is (a) equal to, or (b) less than, or (c) greater than the ratio $w_3 : w_4$; and since r is positive, these three cases arise when we have (a) $p = q$, (b) $p < q$, (c) $p > q$.

7.6 Measuring Probabilities of Indifferent Propositions

Using (7-3) and (7-4) we can now determine the probability of any indifferent proposition X to any finite degree of accuracy.

The cases where X has probability 0 or 1 are recognizable via (7-3) as the ones where X is null or $\overline{X}$ is. Then let us assume that neither X nor $\overline{X}$ is null, as well as that X is ranked with T.

To determine the probability of X

$$x = prob\ X$$

to an accuracy of one part in $M = 2^m$, apply the splitting procedure m times to G and m times to $\overline{G}$ (as in example 5, where $m = 10$ and so $M = 1{,}024$). This yields M equiprobable incompatible propositions G_n ($n = 1, \ldots, M$) whose disjunction is G and which are ranked with each other and with G, and M equiprobable incompatible propositions H_n whose disjunction is $\overline{G}$ and which are ranked with each other and with $\overline{G}$. For each n the proposition

$$I_n = G_n \vee H_n$$

will be indifferent and have probability $1/M$, and the proposition

$$S(n) = I_1 \vee \ldots \vee I_n$$

will be indifferent and have probability n/M. To round things out, define

$$S(0) = F$$

The propositions $S(n)$ form an M-step probability scale from *prob* $S(0) = 0$ to *prob* $S(M) = 1$ against which prob X can be measured. This is a matter of using (7-4), say with $A = X$ and with $B = \mathrm{S}(n)$, in order to determine which of the cases

$$\text{(a) } x = n/M\ , \qquad \text{(b) } x < n/m\ , \qquad \text{(c) } x > n/M$$

holds, for $n = 1, \ldots, M - 1$. (Since neither X nor $\overline{X}$ is null, we know that case (c) holds for $n = 0$ and case (b) holds for $n = M$.) If case (a) holds for some n then we have determined

$$x = \frac{n}{M}$$

exactly. And if case (a) holds for no n, there will be a pair of successive numbers, n, $n + 1$, where case (c) holds for the first and case (b) holds for the second—so that we shall have determined that *prob* X lies in the interval from *prob* $S(n)$ to *prob* $S(n + 1)$:

$$\frac{n}{M} < x < \frac{n}{M} + \frac{1}{M}$$

It only remains to verify that the conditions of applicability of (7-4) are met whenever X is indifferent and neither X nor $\overline{X}$ is null, provided $n = 1, \ldots, M - 1$.

Note that as $\overline{X}$ is not null and X is indifferent, $\overline{X}$ must be indifferent too, for the conditions

$$des\ X = des\ T, \quad des\ T = x\ des\ X + (1 - x)\ des\ \overline{X}$$

imply that $des\ \overline{X} = des\ T$.

As X and $\overline{X}$ are indifferent, and G and $\overline{G}$ are not indifferent, and none of X, $\overline{X}$, G, $\overline{G}$ is null, one can show that at most one of GX, $G\overline{X}$, $\overline{G}X$, $\overline{G}\overline{X}$ is null, i.e., that no distinct pair of those four propositions can have a null disjunction without violating one of the foregoing conditions on X and G. There are six such pairs:

Pair	Disjunction	Condition Violated if the Disjunction is Null
GX, $G\overline{X}$	G	G is not null
GX, $\overline{G}X$	X	X is not null
GX, $\overline{G}\overline{X}$	$GX \vee \overline{G}\overline{X}$	G is not ranked with $\overline{X}$
$G\overline{X}$, $\overline{G}X$	$G\overline{X} \vee \overline{G}X$	G is not ranked with X
$G\overline{X}$, $\overline{G}\overline{X}$	$\overline{X}$	$\overline{X}$ is not null
$\overline{G}X$, $\overline{G}\overline{X}$	$\overline{G}$	$\overline{G}$ is not null

Now there are five cases to consider, in showing that the conditions of applicability of (7-4) are met whenever X and $\overline{X}$ are not null, X is indifferent, and $n = 1, \ldots, M - 1$. In each case, the problem is to identify a suitable test proposition, C, that is neither null nor indifferent and is incompatible with $S(n)$ and with the proposition with which its probability is to be compared (i.e., either X itself or some indifferent proposition that has the same probability as X).

Case 1. GX is null (and so none of $G\overline{X}$, $\overline{G}X$, $\overline{G}\overline{X}$ are). Then $x = prob\ X = prob\ GX + prob\ \overline{G}X = prob\ \overline{G}X$, where $\overline{G}X$ is indifferent because $X = GX \vee \overline{G}X$ is, and $prob\ GX = 0 \neq prob\ \overline{G}X$. Now in (7-4), $A = \overline{G}X$, $B = S(n)$, and if we set $C = G_{n+1}\overline{X}$, the conditions are met, i.e., $prob\ C = prob\ G_{n+1} = (prob\ G)/M$ where $prob\ G \neq 0$, so that C is not null; C is ranked with G and so is not indifferent; and $CA = G_{n+1}\overline{X}\overline{G}X = F$ and $CB = G_{n+1}\overline{X}S(n) = F$ since $G_{n+1}S(n) = F$.

Case 2. GX is null. As $\overline{X}$ is indifferent, the same argument can be used as in case 1, with "X" and "$\overline{X}$" interchanged.

Case 3. $\overline{G}X$ is null. Here the same argument can be used as in case 1 if we imagine that the notation has been changed in the definition of I_n and $S(n)$; "G" and "$\overline{G}$" are interchanged, "G_i" and "H_i" are interchanged, but the propositions $S(n)$ turn out to be just as they were in the original notation.

Case 4. $\overline{G}\overline{X}$ is null. Here the same argument can be used as in case 2, if we imagine that the further notational changes noted in case 3 have been made.

Case 5. None of GX, $G\overline{X}$, $\overline{G}X$, $\overline{G}\overline{X}$ is null. Here there are two subcases, as follows.

(a) $G\overline{X}$ is not indifferent. With $A = X$ and $B = S(n)$ in (7-4), set $C = G_M\overline{X}$, where the numbering of the M slices of G has been arranged so that $G_M\overline{X}$ is not null and not indifferent. (Since $G\overline{X}$ is not null, *some* slice of G must have a conjunction with $\overline{X}$ that is not null. And since $G\overline{X}$ is not indifferent, not all of its non-null pieces of form $G_i\overline{X}$ can be indifferent.) Clearly, $AC = BC = F$ when C is so chosen. Then the conditions of applicability of (7-4) are met.

(b) $G\overline{X}$ is indifferent. The GX is not, since $G = GX \vee G\overline{X}$ is good, where neither disjunct is null. Now repeat the argument of case 5(a) with

"X" and "$\overline{X}$" interchanged, to show that *prob* $\overline{X}$ can be determined to within an accuracy of one part in M. It follows that $x = 1 - prob\,\overline{X}$ can also be determined to that accuracy.

This completes the proof.

What we have established in this chapter, and will have much use for in the uniqueness proof of chapter 8, can be summarized as follows.

(7-5) If the existence, closure, G, and splitting conditions are satisfied, then (a) the ratio of probabilities of any two propositions that are ranked together is uniquely determined by the preference ranking, and therefore (b) the probabilities of all indifferent propositions are uniquely determined by the preference ranking.

7.7 Problems

1

Prove that A is more probable than G if the preference ranking is as follows.

$$\overline{A}$$

$$\overline{B}$$

$$T$$

$$A$$

$$B$$

2

Suppose that A and B are both bad, that A is preferred to B, and that A is more probable than B. Does it follow that the preference ranking is as shown in problem 1?

3

Why was A required to be good or bad in the splitting condition?

4

What is the probability of A when the preference ranking is as follows?

$$\overline{A}$$

$$A, T$$

5

Apply the splitting condition twice to G, to get four incompatible equiprobable slices of G:

$$G_1, G_2, G_3, G_4\ .$$

In addition, get four incompatible equiprobable slices of $\overline{G}$,

$$H_1, H_2, H_3, H_4\ ,$$

in the same way. Define

$$I_1 = G_1 \vee H_1 \quad I_2 = G_2 \vee H_2 \quad I_3 = G_3 \vee H_3 \quad I_4 = G_4 \vee H_4$$

(a) What is the probability of

$$I_1 \vee I_2 \vee I_3\ ?$$

(b) What is the preference ranking of the following propositions?

$$G, \overline{G}, I_1, I_1 \vee G_2, I_1 \vee H_2, I_1 \vee G_2 \vee G_3, I_1 \vee I_2 \vee H_3,$$
$$I_1 \vee I_2 \vee G_3 \vee G_4, I_1 \vee I_2 \vee I_3 \vee H_4, I_1 \vee G_1$$

6

The condition that

(a) $$prob\ X = des\ X$$

for all propositions X, in the preference ranking, would represent an extreme form of wishful thinking. The condition that for all propositions X, Y in the preference ranking

(b) $X \vee Y$ is ranked with X if X is ranked no higher than Y

would represent an extreme form of fearful thinking. Show that each of these conditions is incompatible with the assumption that the existence, closure, G, and splitting conditions all hold.

7

A and B are good propositions that are ranked together. The splitting procedure has been applied once to A and once to B, yielding slices A_1 and A_2 of A, and slices B_1 and B_2 of B. Write out the preference ranking of the propositions

$$A, B, \overline{A}, \overline{B}, A_1, A_2, \overline{A_1}, \overline{A_2}, B_1, B_2, \overline{B_1}, \overline{B_2}$$

on each of the following assumptions:

(a) $prob\ A = prob\ B$

(b) $prob\ A = 2\ prob\ B$

(c) $prob\ A = \frac{1}{2}\ prob\ B$

(d) $prob\ A = \frac{3}{4}\ prob\ B$

8

Show that if the existence, closure, and G conditions are satisfied and if the ranking of A, B and $A \vee B$ is

$$A$$

$$A \vee B$$

$$B$$

whenever $AB = F$ and A is preferred to B, then none of the propositions in the preference ranking have probability 0.

8

Uniqueness

We can now prove that under the conditions described in section 7.1 the transformations discussed in chapter 6 comprise all of the preference-preserving probability and desirability transformations.

> UNIQUENESS THEOREM. Suppose that the closure condition, the G condition, and the splitting condition are satisfied, and that the existence condition is satisfied in the case of two pairs of assignments: *prob, des* and *PROB, DES*. Then there are real numbers a, b, c, d that simultaneously satisfy the following five conditions (where (b), (d), and (e) are asserted for all non-null propositions A in the preference ranking).

(8-1)

(a) $ad - bc$ is positive

(b) $c\ des\ A + d$ is positive

(c) $c\ des\ T + d = 1$

(d) $$DES\ A = \frac{a\ des\ A + b}{c\ des\ A + d}$$

(e) $PROB\ A = prob\ A\ (c\ des\ A + d)$

The hypothesis that the existence condition is satisfied in the case of both pairs, *prob, des* and *PROB, DES,* means that the probability and desirability axioms (5-1) and (5-2) hold for *prob, des;* that they continue to hold when "*prob*" and "*des*" are capitalized; and that for any propositions A, B in the preference ranking, these three conditions all hold, or all fail:

$des\ A \leq des\ B$, $\quad DES\ A \leq DES\ B$, $\quad A$ is ranked no higher than B.

In proving the theorem, we shall have much use for the result (7-5) of the discussion in chapter 7, which can now be rephrased as follows.

(8-2) Under the hypotheses of the uniqueness theorem,

(a) *prob A PROB B* = *PROB A prob B* if *A* and *B* are ranked together; and in particular,
(b) *prob A* = *PROB A* if *A* is indifferent.

To see that (8-2) (a) is simply another way of writing (7-5) (a), rewrite (8-2)(a) as

$$\frac{prob\ A}{prob\ B} = \frac{PROB\ A}{PROB\ B},$$

which means that the ratio of the probability of *A* to the probability of *B* is the same in any probability scale that is consistent with the preference ranking. In other words, the ratio is uniquely determined by the preference ranking. Condition (8-2) (b) is the special case of (8-2) (a) in which *B* = *T*, so that *prob B* = *PROB B* = 1.

8.1 Uniqueness of Probabilities

To prove the uniqueness theorem, it suffices to show that real numbers *a, b, c, d* exist which satisfy the first four of conditions (8-1) (a) through (e), including the condition that *DES* be related to *des* as in condition (d): from that it follows that *PROB* and *prob* will be related as in condition (e). To verify this claim, recall that desirabilities determine probabilities as in section 5.9. There are two cases to consider.

In case *A* is indifferent (and so is ranked with *T*), we have

$$\begin{aligned} PROB\ A &= prob\ A && \text{by (8-2) (b)}, \\ &= prob\ A(c\ des\ T + d) && \text{by (6-3) (c)}, \\ &= prob\ A(c\ des\ A + d) && \text{as } A \text{ is ranked with } T. \end{aligned}$$

Then equation (8-1) (e) is satisfied by any indifferent *A*.

In case *A* is not indifferent, i.e., not ranked with *T*, it will not be ranked with $\overline{A}$, either. (If *X* and $\overline{X}$ are ranked together, (5-3) requires them both to be ranked with *T*.) Then by (5-4) we have

$$PROB\ A = \frac{DES\ T - DES\ \overline{A}}{DES\ A - DES\ \overline{A}} = \frac{Z - Y}{X - Y},$$

$$prob\ A = \frac{des\ T - des\ \overline{A}}{des\ A - des\ \overline{A}} = \frac{z - y}{x - y},$$

where the right-hand forms are in an obvious shorthand (in which "Z" stands for "DES T", etc.). In that same shorthand, (8-1) (d) yields

$$X = \frac{ax + b}{cx + d}, \quad Y = \frac{ay + b}{cy + d}, \quad Z = \frac{az + b}{cz + d}.$$

From these five, together with (8-1) (a)–(c), one can derive the form

$$\frac{PROB\ A}{prob\ A} = cx + d$$

of (8-1) (e) by straightforward algebraic manipulation.

8.2 A Scale of Desirabilities between 0 and 1

To highlight the essentials, let us now focus on the case where the *des* and *DES* scales have a common zero, *T*, and a common unit, *G:*

(8-3) (a) $des\ T = DES\ T = 0$, (b) $des\ G = DES\ G = 1$.

Then in Ramsey's utility theory, or in that of von Neumann and Morgenstern, the two scales would be one and the same. But here, the effect is to fix two of the degrees of freedom, corresponding to a choice of parameter values

$$b = 0 , \quad d = 1 ,$$

without making any special stipulation about the value of the third parameter, i.e., c, or, equivalently, a ($= c + 1$). Then conditions (8-1) (d) and (e) assume the special forms

(8-4) (a) $DES\ A = \dfrac{(c + 1)\ des\ A}{c\ des\ A + 1}$,

(b) $PROB\ A = prob\ A\ (c\ des\ A + 1)$,

and the condition (8-1) (a) that $ad - bc$ be positive becomes simply

$$c > -1 .$$

Much as we constructed probability scales in chapter 7, we now construct a scale of desirabilities for propositions ranked between *T* and *G*, i.e., according to (8-3), for propositions with desirabilities in the unit interval. To construct a scale of a certain degree of fineness, we apply the splitting procedure *m* times to *G*, and *m* times to $\overline{G}$, for some positive integer *m*. This will yield $M = 2^m$ equiprobable slices $G_1, \ldots, G_M$ of *G*, and *M* equiprobable slices $H_1, \ldots, H_m$ of $\overline{G}$. Now we define

$$\begin{array}{ll} G(0) = F & I(0) = T = G \vee \overline{G} \\ G(1) = G_1 & I(1) = G_2 \vee \ldots \vee G_M \vee H_2 \vee \ldots \vee H_M \\ G(2) = G_1 \vee G_2 & I(2) = G_3 \vee \ldots \vee G_M \vee H_3 \vee \ldots \vee H_M \\ \cdot & \cdot \\ \cdot & \cdot \\ \cdot & \cdot \\ G(n) = G_1 \vee G_2 \vee \ldots \vee G_n & I(n) = G_{n\omega 1} \vee \ldots \vee G_M \vee H_{n\omega 1} \vee \ldots \vee H_M \\ \cdot & \cdot \\ \cdot & \cdot \\ \cdot & \cdot \\ G(M) = G & I(M) = F \end{array}$$

The point of all this apparatus is illustrated in figure 8.1 for the case in which the splitting procedure has been applied $m = 3$ times. In general, given (8-3), we have

$$\text{(8-5)} \quad des\,[G(n) \vee I(n) = \frac{des\,G(n)\,prob\,G(n) + des\,I(n)\,prob\,I(n)}{prob\,G(n) + prob\,I(n)}$$

$$= \frac{prob\,G(n)}{prob\,G(n) + prob\,I(n)}$$

$$= \frac{(n/M)prob\,G}{(n/M)prob\,G + (M - n)/M},$$

Fig. 8.1 An eight-step desirability scale. If G has probability 3/5, the desirability of the shaded proposition is 1/2.

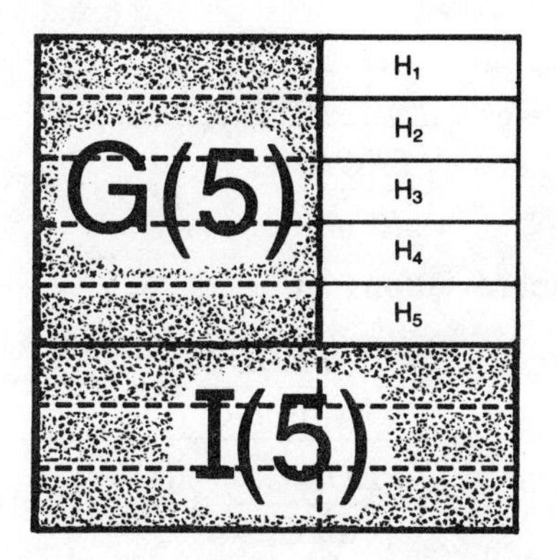

since $G(n)$ is the union of n of the M equiprobable slices of G, and $I(n)$ is the union of $M - n$ of the M equiprobable slices of $G \vee \overline{G}$. Then if *prob* $G = 3/5$ in figure 8.1, we have

$$des\ [G(5) \vee I(5) = \frac{(5/8)(3/5)}{[(5/8)(3/5) + 3/8} = \frac{1}{2}$$

for the desirability of the shaded region.

8.3 Uniqueness of the Scale

But generally speaking, the preference ranking does not determine a precise numerical value for the probability of G. Let us abbreviate (8-5) by setting $des\ [G(n) \vee I(n)] = s$, so that (8-5) becomes

$$s = \frac{r\ prob\ G}{r\ prob\ G + (1 - r)} \tag{8-6}$$

where r is the ratio n/M. If *DES* is a second desirability scale satisfying (8-3) and the conditions specified in the preamble to the uniqueness theorem, we shall have

$$S = DES[G(n) \vee I(n) = \frac{r\ PROB\ G}{r\ PROB\ G + (1 - r)}, \tag{8-7}$$

and we shall not have $S = s$ unless it happens that $PROB\ G = prob\ G$, so that the two scales are in fact one and the same.

If *DES* and *des* are not identical, how are they related? We can answer this question by solving (8-6) for r,

$$r = \frac{s}{prob\ G + (1 - prob\ G)s},$$

substituting this into (8-7), and simplifying, to get

$$S = \frac{\dfrac{PROB\ G}{prob\ G}s}{\left(\dfrac{PROB\ G}{prob\ G} - 1\right)s + 1},$$

or, eliminating the abbreviations "s" and "S,"

$$DES[G(n) \vee I(n) = \frac{\dfrac{PROB\ G}{prob\ G} des\ [G(n) \vee I(n)}{\left(\dfrac{PROB\ G}{prob\ G} - 1\right) des\ [G(n) \vee I(n) + 1}. \tag{8-8}$$

But this is the special case of (8-4) (a) in which A is of form $G(n) \vee I(n)$ and c is determined by the probabilities of G on the two scales,

$$a = c + 1 = \frac{PROB\ G}{prob\ G} \tag{8-9}$$

We have thus proved the uniqueness theorem in the special case in which the proposition A mentioned in the theorem is one of the *scaling propositions*, $A = G(n) \vee I(n)$ for some $n \leq M = 2^m$. The restrictions (8-3) are of no consequence, since where they are not met by the given scales (say, *des'* and *DES'*), we can easily transform them into equivalent scales *des, DES* that do satisfy (8-3). Reasoning as above, we find that indeed there are real numbers

$$a = \frac{PROB\ G}{prob\ G}, \quad b = 0, \quad c = a + 1, \quad d = 1,$$

satisfying (8-1) (a–d). (As observed in section 8.1, (8-1) (e) must then be satisfied as well.) If we now transform *des* and *DES* back to the original scales, *des'* and *DES'*, we find that there are four real numbers (say, a', b', c', and d') that satisfy the analogues of (8-1) (a–d) in which *des'* and *DES'* replace *des* and *DES*.

8.4 Uniqueness of Desirabilities in the Unit Interval

Let us now return to the assumption (8-3) that *des* and *DES* have T as a common zero, and G as a common unit. As we have just seen, no real generality is lost in this way.

The scaling propositions $G(n) \vee I(n)$ have desirabilities

$$s = des\ [G(n) \vee I(n)]$$

that mark $M + 1$ different points from 0 to 1 as n assumes the $M + 1$ values $0, 1, \ldots, M$. By increasing m by 1, i.e., by doubling $M = 2^m$, we separate these by M new points. Then if we consider all the scaling propositions obtained when m ranges over the positive integers, there will be no end of different values that s will assume in the interval from 0 to 1: the "unit interval." But if the scaling propositions are to be used in the way their name promises, as a way of measuring *des A* to any desired accuracy when A lies between T and G in the preference ranking, we must be assured that for any number x in the interior of the unit interval, and any positive error e, no matter how small, x will be flanked above and below by desirabilities s of scaling propositions—desirabilities that differ from x by less than e. The mere existence of infinitely many different rankings of scaling propositions between T and G does not guarantee that, for the desirabilities of those propositions *might* cluster in such a way as to leave bare stretches, e.g., for all we know until we prove otherwise, they might all fall outside the middle third of the unit interval.

To prove that every point in the unit interval is confined as tightly as you please between scaling propositions, we been to modify our notation so as to explicitly indicate both n and M. Thus, we shall write "$G(n,M)$" and "$I(n,M)$" in place of "$G(n)$" and "$I(n)$" as above, so that, e.g., the shaded proposition in figure 8.1 is now $G(5,8) \vee I(5,8)$, with $2^m = M = 8$ indicated explicitly. For any x in the interior of the unit interval and for any $M = 2, 4, 8, \ldots$, there will be a unique n for which the desirabilities of adjacent scaling propositions bracket x in the sense that the inequality

$$\text{(8-10)} \quad des\,[G(n,M) \vee I(n,M) \leq x < des\,[G(n+1,M) \vee I(n+1,M)$$

is satisfied. Let us write

$$\Delta(M,x) = des\,[G(n+1,M) \vee I(n+1,M)] - des\,[G(n,M) \vee I(n,M)]$$

for this particular n.

In these terms, what we want to prove can be expressed as follows.

SCALING LEMMA. For each x in the interior of the unit interval, the differences $\Delta(M,x)$ get and remain smaller than any preassigned positive limit, as M increases without bound.

Proof. For the n in question, set

$$r(M,x) = \frac{n}{M}$$

and notice that for fixed x, $r(M,x)$ is a nondecreasing function of M. By (8-6),

$$\Delta(M,x) = \frac{\left(r + \frac{1}{M}\right) prob\, G}{\left(r + \frac{1}{M}\right) prob\, G + 1 - \left(r + \frac{1}{M}\right)} - \frac{r\, prob\, G}{r\, prob\, G + 1 - r}$$

$$= \frac{prob\, G}{(r\, prob\, G + 1 - r)[prob\, G - 1 + M(r\, prob\, G + 1 - r)}$$

where in this last quotient, the denominator increases without bound as M does, provided $r\, prob\, G + 1 - r$ is not 0, i.e., provided $prob\, G$ is not the (negative) number $(1 - r)/r$—as it cannot be, by (5-1) (a). Then as the numerator is constant, $\Delta(M,x) \longrightarrow 0$ as $M \longrightarrow \infty$, Q.E.D.

Now as we have established the uniqueness result (8-4) (a) for desirabilities of the scaling propositions themselves, and we have found that all points in the unit interval can be approximated with any desired degree of accuracy by desirabilities of scaling propositions, we conclude that the relationship (8-4) (a) between *des A* and *DES A* holds for all *A* between *T* and *G*.

To complete the proof of the uniqueness theorem, we need only prove the same result for propositions *A* that are ranked above *G*, or below *T*. In view of (8-3), this is a matter of proving uniqueness of desirabilities greater than 1, and less than 0.

8.5 Uniqueness of Negative Desirabilities

In case *A* is ranked below *T*, apply the splitting procedure to it one or more times in order to obtain incompatible propositions $A_1, A_2, \ldots, A_M$ which are ranked with *A* and for which we have

$$\text{(8-11)} \qquad prob\ A_1 = prob\ A_2 = \frac{prob\ A}{M} = p$$

$$PROB\ A_1 = PROB\ A_2 = \frac{PROB\ A}{M} = q$$

By choosing *M* large enough, the probability of A_1 and A_1 on either scale can be made so small that both $\overline{A_2}$ and $\overline{A_1 \vee A_2}$ are ranked below *G*, as shown in table 8.1, for, according to (5-6), the desirability of the denial of a proposition can be brought as close to 0 as you please by making the probability of the proposition sufficiently small.

Now by (8-3) we have

$$\begin{aligned} 0 &= des\ (A_1 \vee \overline{A}) \\ &= DES\ (A_1 \vee \overline{A}_1) \\ &= des\ [(A_1 \vee A_2) \vee (\overline{A_1 \vee A_2})] \\ &= DES\ [(A_1 \vee A_2) \vee (\overline{A_1 \vee A_2})] \end{aligned}$$

so that by the desirability axiom, in the notation of (8-11) and of table 8.1,

Table 8.1

Ranking	*des*	*DES*
G	1	1
$\overline{A_1 \vee A_2}$	z	w
$\overline{A}_1$	y	v
T	0	0
$A,\ A_1,\ A_1 \vee A_2$	x	u

we have

(8-12)

$$
\begin{aligned}
&(a)\quad 0 = px + (1 - p)y \\
&(b)\quad\;\; = qu + (1 - q)v \\
&(c)\quad\;\; = 2px + (1 - 2p)z \\
&(d)\quad\;\; = 2qu + (1 - 2q)w\,.
\end{aligned}
$$

Applying the uniqueness result of section 8.4 to the propositions $\overline{A_1 \vee A_2}$ and $\overline{A}_1$ (which have desirabilities in the unit interval) we get two further equations,

$$(e)\; w = \frac{(c + 1)z}{cz + 1}$$

$$(f)\; v = \frac{(c + 1)y}{cy + 1}$$

With some drudgery, these six equations can be solved for u in the form

(8-13) $$u = \frac{(c + 1)x}{cx + 1}$$

(Suggestion: work with (a), (b), and (f) as one group, and with the other three as another. From each group, obtain an equation involving all and only p, q, x, u, and c. Combine these to eliminate q, and simplify to get [8-13].)

This extends the uniqueness result to negative desirabilities.

8.6 Completing the Uniqueness Proof

Having proved the uniqueness theorem for propositions ranked no higher than G, it only remains to treat the case where *des A* and *DES A* are both greater than 1. Here, we split A once to obtain the preference ranking shown in table 8.2, where now we set

$$p = prob\ A_1 = ½ prob\ A$$

$$q = PROB\ A_1 = ½ PROB\ A$$

and "u" through "z" have the significance shown in table 8.2.

Table 8.2

Ranking	*des*	*DES*
A, A_1	x	u
G	1	1
T	0	0
$\overline{A}_1$	y	v
$\overline{A}$	z	w

In this notation, we get the same six equations (8-12) (a–f) as in section 8.5, where the last two are now justified by the fact, proved in section 8.5, that the uniqueness result holds for negative desirabilities. As before, (8-13) follows, showing (according to the interpretation in table 8.2) that the uniqueness theorem holds for desirabilities greater than 1.

This completes the proof of the uniqueness theorem.

8.7 Problems

1

Why is the proposition *A* required to be non-null in the uniqueness theorem?

2

Suppose that the pair *prob, des* and the pair *PROB, DES* both satisfy the existence condition, for a certain preference ranking. Call a sequence of four propositions *V, W, X, Y harmonic* (relative to *des*) if we have

$$\text{(8-14)} \qquad \frac{des\ X - des\ V}{des\ X - des\ W} = -\frac{des\ Y - des\ V}{des\ Y - des\ W}$$

Prove that if (8-14) holds we must also have

$$\frac{DES\ X - DES\ V}{DES\ X - DES\ W} = -\frac{DES\ Y - DES\ V}{DES\ Y - DES\ W}$$

so that in fact, the property of being a harmonic sequence depends only on the preference ranking and is independent of which desirability assignment is used to represent it.

3

Prove that if the sequence *T, G, X, Y* is harmonic, and $des\ T = 0$ and $des\ G = 1$, then

$$\text{(8-15)} \qquad des\ Y = \frac{1}{2 - (1/des\ X)}$$

so that as *X* moves toward the top (or bottom) of the preference ranking, *des Y* moves toward 1/2 from above (or below), and gets as close as you please to 1/2 if the desirability assignment *des* is unbounded above (or below).

4

Prove the correctness of the following observation of Ethan Bolker's.

(8-16) If (a) $AB = BC = AC = F$ and

(b) A and B are ranked together but not with C and
(c) $A \vee C$ is ranked with $B \vee C$ and
(d) the existence condition is satisfied, then
(e) the sequence $A, A \vee C, A \vee B \vee C, C$ is harmonic.

5

Suppose that the existence condition, the closure condition, the G condition, the splitting condition, and the following *unboundedness condition* all hold. It follows that if the pair *prob, des* meet the existence condition, then *des* is unbounded both above and below; and it follows that if $des\ T = 0$ and $des\ G = 1$, then $des\ D = 1/2$, where D is as described in the following unboundedness condition. Why?

(8-17) UNBOUNDEDNESS CONDITION. All propositions D that satisfy conditions (a) and (b) below are ranked together.

(a) Whenever A, B, C are pairwise incompatible propositions of which A and B are ranked with T and C is ranked above $A \vee C$, which in turn is ranked *with* $B \vee C$ and *with* or *above* G, the proposition $A \vee B \vee C$ is ranked above D.

(b) Whenever A, B, C are pairwise incompatible propositions of which A and B are ranked with G and C is ranked below $A \vee C$, which in turn is ranked *with* $B \vee C$ and *with* or *below* T, the proposition $A \vee B \vee C$ is ranked below D.

6

Explain why the following condition deserves its name.

(8-18) BOUNDEDNESS CONDITION. There are propositions D_1 and D_2 which are ranked between T and G and are not ranked together; and clauses (a) and (b) of (8-17) both hold when $D = D_1$, and when $D = D_2$.

8.8 Notes and References

The present uniqueness theorem is a version of a theorem proved by Ethan Bolker in his doctoral dissertation, "Functions Resembling Quotients of Measures" (Harvard University, April 1965). The mathematical part of the dissertation has been published, somewhat condensed, in *Transactions of the American Mathematical Society* 124 (1966): 292–312. The application to decision theory, as in chapter 9 of the dissertation, appears in "A Simultaneous Axiomatization of Utility and Subjective Probability," *Philosophy of Science* 34 (1967): 333–40.

When we met, late in 1963, Bolker had been looking for an application of his mathematical theorems, and I had been trying to work out the mathematics of my version of Bayesianism. Each of us had what the other needed. I had been trying, off and on for over a year, to prove the stronger, false variant of the present uniqueness theorem in which $c = 0$ and $d = 1$, i.e., I had been trying to prove the uniqueness theorem with (8-1) replaced by the conditions

(8-19) (a) $DES\ A = a\ des\ A + b\ ,\ a > 0$

(b) $PROB\ A = prob\ A$

that hold in Ramsey's theory, in von Neumann and Morgenstern's, and in the other theories current at the time. (In those theories, probability really *is* unique, and desirability is unique once the zero and unit are chosen.) It simply had not crossed my mind that *prob* and *des* could be subjected to a broader class of transformations than those satisfying (8-19) without ceasing to represent the same preference ranking, or ceasing to satisfy the probability and desirability axioms. Just before Bolker and I met, Kurt Gödel conjectured the correct form of the uniqueness result, and he suggested a way of using linear algebra to prove it. Unknown to Gödel and to me, Bolker had already proved the theorem using projective geometry. Gödel told me his idea, Bolker showed me his proof, and I didn't really understand either. But I thought I did and thought they came to much the same thing, i.e., something like the proof given in this chapter—a proof that has the virtue of accessibility to people who haven't forgotten all of their high school algebra and haven't learned any more mathematics than that. That's why I use it here.

Bolker proves his uniqueness theorem in two different ways, both in his dissertation and in his 1966 article, and states the theorem without proof in his 1967 article. The hypotheses of the present uniqueness theorem are somewhat weaker than those in Bolker's, where the closure condition and the assumption that *prob* is additive (i.e., (5-1) (c) here) are both stronger. Bolker's closure condition is this: *the propositions in the preference ranking form a complete, atom-free Boolean algebra*. For *prob* he assumes *countable* additivity. (The new terminology is explained in chapter 9.) Another difference: the splitting condition, which does follow from Bolker's hypotheses, plays no role in his proofs. But of course, to describe Bolker's theorem from the present perspective is to distort the history of the matter: the present version is simply my attempt to present Bolker's theorem in a mathematically undemanding way.

9

Existence: Bolker's Axioms

In chapters 6 and 7 we simply assumed the existence of probability and desirability functions that represent the agent's preferences. The assumption is fairly plausible, at least as an ideal, for as the epigraph to chapter 1 testifies, the view that preferences ought to be representable in that way has been urged by authors of "How-to-Think" books for over three centuries.

The *existence problem* is the problem of formulating intelligible non-quantitative conditions on preference rankings which imply the existence of quantitative *prob* and *des* functions that represent those rankings: it is the problem of axiomatizing the theory of preference. The existence problem is mathematically deeper than the "uniqueness" problem of specifying to what extent such *prob* and *des* functions are unique when they exist. For the present theory of preference, both problems were solved by Bolker in 1964. In chapter 8 we were able to present a variant of his uniqueness theorem and to provide an elementary proof for it. But in this chapter we can only explain the existence theorem without proof: the needed mathematics lie far beyond the high-school level that suffices for the rest of this book.

9.1 Preference-or-Indifference as a Primitive

Axiom 1 makes explicit the assumptions that are smuggled in when we speak metaphorically of preference as a matter of height in a ranking where higher is better and where the agent is indifferent between prospects at the same level. It makes for economy to use preference-or-indifference, $\succsim$, as the basic relation. Thus, in terms of the spatial metaphor, "$A \succsim B$" means that A is at least as high as B in the agent's preference ordering.

AXIOM 1. The relation $\succsim$ is transitive and connected.

Here are definitions of the new terms.

Transitivity: if $A \succsim B$ and $B \succsim C$ then $A \succsim C$.

(If A is at least as high as B, which is at least as high as C, then A is at least as high as C.) Connexity means that any two prospects that appear anywhere in the ranking are related by $\succsim$ in one direction or the other. Here, the talk of "appearing in the ranking" is to be understood as membership in the *field* of the relation $\succsim$, as follows.

> The *field* of a two-place relation (e.g., $\succsim$) is the set of all those items A that bear that relation to something ($A \succsim B$) or to which something bears that relation ($B \succsim A$).

We can now define

Connexity: $\succsim$ connects its field, i.e., if A and B are both in the field of the relation, then $A \succsim B$ or $B \succsim A$ (or both).

In terms of preference-or-indifference, we define preference ($\succ$) and indifference ($\approx$) as follows.

Preference: $A \succ B$ if and only if $A \succsim B$ but not $B \succsim A$.
Indifference: $A \approx B$ if and only if $A \succsim B$ and $B \succsim A$.

It is now straightforward to prove (using axiom 1) that preference is irreflexive (not $A \succ A$), asymmetric (if $A \succ B$, then not $B \succ A$), and transitive (if $A \succ B$ and $B \succ C$, then $A \succ C$); that indifference is an equivalence relation (i.e., symmetric, reflexive, and transitive, where "symmetric" means that if $A \approx B$ then $B \approx A$); and that the law of trichotomy holds, i.e., if A and B are in the field of the relation $\succsim$, then exactly one of these three conditions holds: $A \succ B$, $A \approx B$, $B \succ A$.

9.2 Prospects as Propositions

Axiom 1 would be no different if we were dealing with Ramsey's sort of theory, or that of von Neumann and Morgenstern. The distinguishing feature of the present theory is the identification of prospects (the things in the field of the relation $\succsim$) as propositions, i.e., items to which the logical operations of conjunction, disjunction, and denial can be applied, and between which such logical relations as implication can hold. Axiom 2 expresses this identification, together with two further assumptions ("complete, atomless") that will be explained in section 9.4.

AXIOM 2. The field of the relation $\succsim$ is a complete, atomless Boolean algebra of propositions from which the impossible proposition F has been removed.

By a Boolean algebra of propositions is meant a set (the propositions) containing at least the two distinct items T, F, and containing, along with each item A, the item $\bar{A}$, and along with each pair of items A, B, the items AB and $A \vee B$. Finally, the following equations must hold for all A, B, C in the set.

$$AT = A = A \vee F$$

$$A\bar{A} = F, \quad A \vee \bar{A} = T$$

$$AB = BA\,, \quad A \vee B = B \vee A$$

$$A(B \vee C) = AB \vee AC\,, \quad A \vee (BC) = (A \vee B)(A \vee C)$$

Axiom 2 makes rather strong assumptions, but they are needed in order to tease out of the agent's preferences the information we need about his numerical probability and desirability functions. Thus, whenever A and B are nonnecessary propositions that appear in the preference ranking, their denials $\bar{A}$, $\bar{B}$ are required to appear there as well. And by connexity (axiom 1), the agent must have a definite attitude of preference or indifference between A and B and between $\bar{A}$ and $\bar{B}$. Thus, as in example 3 of chapter 7, various numerical equalities and inequalities may be discovered between *prob* A and *prob* B.

9.3 Averaging, Nullity, and Impartiality

The next axiom asserts that disjunction is an averaging operation, as far as preference is concerned: the disjunction of any two incompatible prospects lies somewhere between them in the preference ranking. In axiom 3 it is to be understood that A and B belong to the field of the relation $\succsim$.

AXIOM 3: If $AB = F$ then
(a) if $A \succ B$, then $A \succ A \vee B$ and $A \vee B \succ B$, and
(b) if $A \approx B$, then $A \approx A \vee B$ and $A \vee B \approx B$.

Thus, by (a), if A is preferred to $\bar{A}$, then A is preferred to T, and T is preferred to $\bar{A}$. But by example 2 of chapter 7, this means that neither A nor $\bar{A}$ has probability 0.

In fact, just as axiom 2 excludes the proposition F from the preference ranking, so axiom 3 excludes all other propositions of probability 0 from the ranking—unless the preference ranking is so impoverished that the necessary test propositions (section 7.5) are lacking. The thought behind the exclusion of F is that the agent cannot be expected to have an attitude of preference concerning impossibilities. A similar thought motivates the exclusion of what the agent regards as practical impossibilities: propositions that he would bet against at any odds.

In chapter 7, we tested equiprobability of incompatible propositions A, B that were ranked together by using a test proposition C, incompatible with A and with B and not ranked with them: the test showed equiprobability in case the disjunctions $A \vee C$ and $B \vee C$ were ranked together. Axiom 4 ("impartiality") stipulates that the choice of different test propositions C cannot make the test yield different results. It is to be understood that A, B, and C are in the field of the relation $\succsim$.

AXIOM 4. Given $AB = F$ and $A \approx B$, if $A \vee C \approx B \vee C$ for some C where $AC = BC = F$ and not $C \approx A$, then $A \vee C \approx B \vee C$ for every such C.

9.4 Completeness, Atomlessness, Continuity

The impartiality axiom is obviously true, if the agent's preferences are represented by a pair *prob, des* as in the existence assumption (section 7.1), and it plays an essential role in Bolker's proof of his existence theorem (in effect, in his derivation of the existence assumption from the axioms). But it is not the sort of assumption that is particularly plausible simply because we are taking prospects to be propositions. The axiom is there because we need it, and it is justified by our antecedent belief in the plausibility of the result we mean to deduce from it.

The remaining assumptions are also there because they are needed in the existence proof; but they are far from being consequences of the existence assumption. Rather, they are very strong idealizations, the point of which is to ensure that the agent's preference ranking is rich enough, and organized in a sufficiently tractable way, to allow us to see in it a model of an interval of real numbers. This is how it goes.

We speak of the class of all prospects that are ranked with a given prospect as the *value* of that proposition. We denote the value of A by "*val* A." Thus, B is a member of *val* A if and only if $A \approx B$. By reflexivity of indifference, A belongs to *val* A. It is not hard to see that the reflexivity of indifference implies that *val* A and *val* B are either completely disjoint, or identical: if *val* A and *val* B have even one proposition in common, then *val* A = *val* B. It is these values that will imitate the real numbers. Like the real numbers, values must be continuous (axiom 5).

The continuity condition requires that the implication relation and the preference relation fit together in a certain way. Let us write "$A \models B$" to indicate that A implies B (i.e., that $AB = A$). Observe that implication is a *partial ordering*, i.e., it is reflexive ($A \models A$), transitive (if $A \models B$ and $B \models C$, then $A \models C$), and antisymmetric (if $A \models B$ and $B \models A$, then $A = B$). But it is not generally a *linear ordering*, i.e., a connected ($A \models B$ or $B \models A$) partial ordering. Still, there are sets of propositions (generally smaller than the whole field of the relation $\models$) that are linearly ordered by $\models$. Such sets are called "implication chains."

The *supremum* of a set of propositions is a proposition S that is implied by every proposition in the set (so that S is an upper bound of the set) and implies every other upper bound (so that S is the least of the upper bounds of that set). Similarly, the *infimum* of a set of propositions is a proposition I that implies every proposition in the set (so that I is a lower bound of the set) and is implied by every other lower bound of the set: a greatest lower bound. We can now explain the meaning of the requirement in axiom 2 that the Boolean algebra be complete:

> A *complete* Boolean algebra is one in which every set of propositions has both a supremum and an infimum.

It is clear that suprema and infima are unique, when they exist. The completeness requirement ensures that they exist.

An *atom* is a proposition other than F that is implied by itself and by F, but by no other propositions.

> An *atomless* Boolean algebra is one in which F is the only proposition that is implied only by itself and F.

Now we can state the continuity axiom:

> AXIOM 5. Whenever the supremum (or infimum) of a chain of prospects lies in a preference interval, all members of the chain after (or before) a certain point lie in that interval.

To be more explicit: Suppose that $A \succ S \succ B$ (or that $A \succ I \succ B$), where S and I are the supremum and the infimum of a chain of prospects. Then there will be a member C of the chain with this characteristic: if D is a member of the chain that is implied by C (or that C implies), then $A \succ D \succ B$.

If we write "*val* $A >$ *val* B" when $A \succ B$, then we can see the effect of axiom 5 as assuring us that the trajectory of *val* D through the values will simulate the trajectory of *des* D through the real numbers. For desirability, continuity comes to this: If S and I are the supremum and the infimum of a chain of prospects, and *des* S (or *des* I) lies in a certain interval of real numbers, then for all D after (or before) a certain point C in the chain, *des* D lies in that interval. Axiom 5 allows us to replace "*des*" by "*val,*" here, and to replace "real numbers" by "values."

9.5 Notes and References

Of the three works of Bolker's cited in section 8.8, this exposition of his axioms is closest to the 1967 "simultaneous axiomatization" paper, for which this may serve as a companion piece. Comparison of these with chapter 9 of his dissertation will indicate the connection between his axiomatization and his representation theorem for functions resembling quotients of measures. For the theorem itself, see the earlier chapters of the dissertation, or the 1966 *Transactions* article. For further discussion of the axiomatization, see my "Frameworks for Preference" and his "Remarks on 'Subjective Expected Utility for Conditional Primitives,' " in Balch, McFadden, and Wu, *Essays on Economic Behavior under Uncertainty* (cited in section 1.8), pp. 74–82.

There is a different sort of axiomatization, turgid with tensors, by Zoltan Domotor: "Axiomatization of Jeffrey Utilities," *Synthese* 39 (1978): 165–210. There the representation yields functions *prob, des* that take their values in fields of nonstandard real numbers, e.g., so that propositions can have infinitesimal probabilties. This relieves the pressures that lead Bolker to require completeness, continuity, and atomlessness of the Boolean algebra, but at a certain cost: see the decoding on page 173 of the equation in the projection axiom, J_2 (p. 175). The attractive feature of Domotor's axioms is that they are necessary, as well as sufficient, for existence of representative pairs *prob, des*. Of course, since the values of these functions need not be real, the problem so solved is a different one from Bolker's.

10

Boundedness; Causality

From its inception, Bayesian desirability theory has been closely tied to the notion that desirabilities are (1) finite and, in fact, (2) bounded. The finiteness condition means that $+\infty$ and $-\infty$ cannot be the desirabilities of any propositions: thus, Pascal's Wager (see problem 7 of section 1.6) is outside the scope of the theory. And the boundedness condition means that there must be numbers s and i which serve as upper and lower bounds on the attainable desirabilities: for all propositions X in the agent's preference ranking we have $i \leqslant des\ X \leqslant s$, where s and i are constants for the particular agent at the time in question.

Example 1: The St. Petersburg game

The player tosses a coin repeatedly until the tail turns up on, say, the nth toss, at which point the game ends, and the player is paid 2^n dollars. For each positive integer n, let L_n be the proposition that the game goes on for exactly n tosses. Thus, L_3 is true if and only if the outcomes of the first three tosses are head, head, tail. Presumably, the player takes the probabilities of the various possible lengths of the game to be

$$prob\ L_1 = 1/2$$

$$prob\ L_2 = 1/4$$

$$prob\ L_3 = 1/8$$

$$\vdots$$

$$prob\ L_n = 1/2^n$$

If the player takes the numerical desirability of receiving N dollars to be N units of desirability, then we have

$$des\ L_1 = 2$$
$$des\ L_2 = 4$$
$$des\ L_3 = 8$$
$$\vdots$$
$$des\ L_n = 2^n$$

Suppose that the agent believes he is about to play the St. Petersburg game. Since the propositions L_1, L_2, L_3, etc. are pairwise incompatible in the sense that we have $L_mL_n = F$ if $m \neq n$, and since they are exhaustive in the sense that we have

$$L_1 \vee L_2 \vee L_3 \ldots = T,$$

it appears that the expected desirability of the situation that the player finds himself in at the beginning of the game is

$$des\ T = prob\ L_1\ des\ L_1 + prob\ L_2\ des\ L_2 + prob\ L_3\ des\ L_3 + \ldots \quad (10\text{-}1)$$

or

$$\begin{aligned} des\ T &= (1/2)(2) + (1/4)(4) + (1/8)(8) + \ldots \\ &= 1 + 1 + 1 + \ldots \\ &= \infty \end{aligned}$$

Then in this case, the boundedness condition and the finiteness condition are both violated: the sequence

$$des\ L_1,\ des\ L_2,\ des\ L_3, \ldots \quad (10\text{-}2)$$

or

$$2, 4, 8, \ldots$$

is unbounded; and, furthermore, there is a proposition T, of which the desirability is infinite.

10.1 The St. Petersburg Paradox

The mere fact that the sequence (10-2) is unbounded does not imply that *des* T is infinite.

Example 2
Suppose that

$$des\ L_n = n$$

for each positive integer n. Then equation (10-1) would yield

$$des\ T = (1/2)(1) + (1/4)(2) + (1/8)(3) + \ldots = 2$$

which is a finite number in spite of the fact that the desirabilities of the L's,

$$1, 2, 3, \ldots,$$

form an unbounded sequence.

Daniel Bernoulli argued in 1738 that as long as the sequence (10-2) is unbounded, a variant of the St. Petersburg game can be devised, in which *des T* is again infinite. The device consists in adjusting the payoffs so that if the game lasts n tosses, the player is paid an amount of money of which the desirability is at least 2^n.

Example 3

If the desirabilities of the various payoffs are as in example 2, form a new game in which the player receives four dollars if the game ends on the first toss, sixteen dollars if the game ends on the second toss, \$256 if the game ends on the third toss, and, in general, $2^{(2^n)}$ dollars if the game ends on the nth toss. Then since the desirability of 2^N dollars is N, the desirability of the payoff in case the game ends on the nth toss is 2^n, and we can conclude here as in example 1 that *des T* is infinite.

This gambit is blocked if the desirability of money is bounded above: if there is a finite number s such that for each positive integer n, the prospect of being about to receive n dollars has desirability less than s. However, the assumption that the desirability of money is bounded above does not imply that desirabilities in general are bounded above: there might be things that money cannot buy that have arbitrarily high desirabilities. Then as long as we suppose (as in Ramsey's theory) that given any two consequences and any proposition, there is a gamble on that proposition in which the actor gets one of the consequences if the proposition is true and the other if the proposition is false, the St. Petersburg game might be played with nonmonetary payoffs; and if there is no finite upper bound on the desirabilities of nonmonetary payoffs, the situation in which the necessary proposition has infinite desirability will arise whenever we conceive the agent to be faced with the prospect of a game of the St. Petersburg type, in which for each n the prize associated with L_n has desirability at least 2^n.

The St. Petersburg paradox is the argument that, given an infinite sequence of consequences whose desirabilities are finite but have no finite upper bound, we can use the St. Petersburg construction to prove the existence of a gamble of infinite desirability. The argument shows that in theories like Ramsey's, unboundedness implies infinity. It would be perfectly satisfactory for there to be no finite upper bound on the desirabilities of the consequences and gambles in which the agent is interested, as long as all desirabilities were finite. However, if the agent is supposed to be concerned with all possible gambles among the consequences that concern him, there will be a gamble which has infinite desirability if there are consequences of finite but arbitrarily high desirability.

Example 4: De contemptu mundi

To see that Bayesian decision theory (in Ramsey's form or in the form presented in this chapter) is inadequate if the agent is taken to attribute infinite desirability to some consequence or to some proposition of positive probability, consider the situation of a man who attributes desirability ∞ to the prospect of heaven, attributes finite desirability to all worldly prospects, and believes that he can make the prospect of heaven more or less likely by performing one act or another. If the agent takes act 1 to bestow probability .99 on the prospect of heaven, and he takes act 2 to bestow probability .01 on that prospect, and if he takes all other consequences of the acts to have finite desirabilities, it seems clear that the agent would and should strongly prefer act 1 over act 2. On the other hand, in the Bayesian account of the matter, the agent is taken to rank the two acts together at the top of his preference scale, since each of them has infinite expected desirability; for we have

$$.99 \times \infty = .01 \times \infty = \infty$$

and if x and y are finite,

$$\infty + x = \infty + y = \infty$$

The case is even more difficult in which the agent assigns desirability $-\infty$ to the prospect of hell and takes an act to bestow a positive probability p on the prospect of heaven and also to bestow a positive probability q on the prospect of hell; for since we have

$$p \times \infty = \infty$$

and

$$q \times -\infty = -\infty$$

the expression for the expected desirability of such an act would either have the form

$$\infty + -\infty$$

(if heaven and hell are the only possible consequences) or have the form

$$\infty + -\infty + rx$$

(if there are other possible consequences, of probability r and expected desirability x); and in either case the expected desirability of the act is indeterminate.

10.2 Resolving the Paradox

The case in which (1) desirabilities are (finite but) unbounded both above and below plays a special role in our theory: then, and only then, the agent's probability assignment is completely determined by his preference ranking, and his desirability assignment is determined by his preference ranking once two arbitrary assignments have been made, say *des* $T = 0$ and *des* $G = 1$. And in our theory, the cases in which (2) desirabilities are bounded above but not below, or (3) below but not above, are not essentially different from the case which is mandatory in Ramsey's theory, where (4) desirabilities are bounded both above and below; for any preference ranking which is describable by a desirability assignment of type (4) can be described equally well by an assignment of type (3) or type (2), and conversely, any assignment of type (2) or type (3) can be changed by a perspective transformation into an equivalent assignment of type (4).

However, our theory, like Ramsey's, is incompatible with the existence of a proposition of positive probability to which the agent assigns infinite positive or negative desirability. Therefore it is essential that we avoid the St. Petersburg paradox: it is essential that we block the argument that seems to lead from the existence of a sequence of propositions of finite but unbounded desirability to the existence of a proposition of positive probability and infinite desirability. In fact, we can do this very easily, because gambles play no special role in our theory. Put briefly and crudely, our rebuttal of the St. Petersburg paradox consists in the remark that anyone who offers to let the agent play the St. Petersburg game is a liar, for he is pretending to have an indefinitely large bank.

Example 5
Since 2^{10} is 1,024 we have

$$2^{10n} = (2^{10})^n = (1{,}024)^n$$

which is slightly greater than

$$(1{,}000)^n = (10^3)^n = 10^{3n} .$$

Then for small n, 2^{10n} can be expressed approximately in the decimal notation by writing 1 followed by $3n$ zeros. Now in the St. Petersburg game as described in example 1, the payoffs are approximately

$1,000, $1,000,000, $1,000,000,000, $1,000,000,000,000, . . .

if the tail turns up first on toss number

10, 20, 30, 40, . . .

Although these possibilities are very improbable, they *are* possible, and therefore an honest offer to allow the agent to play the St. Petersburg game ought to be backed up by ability to raise such amounts as one million billions of dollars in the highly unlikely event that the tail turns up first on the fiftieth toss, or one billion billions of dollars in the even more unlikely event that the tail turns up first on the sixtieth toss. But there is not that much money in the world.

To put the matter more carefully, we should notice that it might be possible for a nation to allow the agent to play the St. Petersburg game, but that in that case there would clearly be a finite upper bound on the desirabilities of the possible payoffs.

Example 6

In order to increase revenue without raising taxes, the Congress establishes a national lottery for millionaires: for a payment of $1,000,000 one may play the St. Petersburg game. Payoffs are made in specially printed bills, so that if tail turns up first on the sixtieth toss, the Treasury department delivers to the winner a crisp new billion billion dollar bill. Due to the resulting inflation, the marginal desirabilities of such high payoffs would presumably be low enough to make the prospect of playing the game have finite expected desirability.

Then it is to be supposed that the prospect of playing the St. Petersburg game as defined in example 1 or as modified in example 3, is either illusory (example 5) or has finite expected desirability (example 6). Moreover, similar remarks apply to versions of the game in which the payoffs are things that money cannot buy. In particular, the following modifications of the game of example 1 represent situations that cannot in fact be set up.

Example 7

Let L_n in example 1 be the proposition that the agent remains alive and in good health for 2^n years, for each $n = 1, 2, 3, \ldots$

Example 8

Let L_n in example 1 be the proposition that the agent is tortured for 2^n hours, for each $n = 1, 2, 3, \ldots$

10.3 Gambles as Causal Relationships

The theory of preference that has been developed in the preceding six chapters makes no explicit use of the notion of a gamble. Indeed, we have sometimes spoken of the necessary proposition as a gamble on (say) the proposition G, in which the possible gain is G and the possible loss is $\overline{G}$. We have spoken in this way for vividness, and to emphasize that to the extent that it makes sense to speak of the present theory as dealing with gambles, the gambles are of a very restricted sort, in which the consequences which serve as the prize and as the loss are the truth and falsity of the very proposition on which the agent is said to be gambling.

Example 9

Suppose that G is the proposition that the agent will get a raise next week. Then we have $T = G \vee \overline{G}$, and T can be construed as a gamble on G,

$$A \text{ if } G,\ B \text{ if not}$$

where in fact $A = G =$ get a raise next week and $B = \overline{G} =$ get no raise next week. The estimated desirability of the gamble is

$$prob\ G\ des\ A + prob\ \overline{G}\ des\ B$$

or

$$prob\ G\ des\ G + prob\ \overline{G}\ des\ \overline{G}$$

which is simply *des* T.

In general, a gamble of form

$$A \text{ if } C,\ B \text{ if not}$$

exists if there is a causal relationship between C, A, and B, in virtue of which A will happen if C does, and B will happen if C does not. If I offer to bet you a dollar at even money that C will happen, and you accept the bet, we have set up a causal relationship between C, A, and B, where A is the proposition that

you pay me $1 at the time we learn whether C is true or false,

and B is the proposition that

I pay you $1 at the time we learn whether C is true or false.

The basis of the causal relationship lies in our good faith and in our ability to produce the required cash at the time in question. It is a relationship that we brought into being of our own free will. Nonetheless, it is a genuine causal relationship: as genuine as the relationship between the proposition that the gas tank is empty and the proposition that the car does not go.

10.4 Our Theory Is Noncausal

The theory developed in the preceding six chapters deals only with such causal relationships as the agent believes (rightly or wrongly) actually obtain among the propositions in his preference ranking. We do not need to know how he would revise his preference ranking if he believed in the existence of further causal relations which may be in serious conflict with the relationships he does believe in. Thus, in Ramsey's theory, if the preference ranking contains the consequences (A) that

there will be a thermonuclear war next week

and (B) that

there will be fine weather next week

and if it contains any gambles on the proposition (C) that

this coin lands head up,

then it must also contain the gamble

A if *C*, *B* if not.

It must contain the gamble

> There will be a thermonuclear war next week if this coin lands head up, and there will be fine weather next week if not.

However, for the agent to consider that this gamble might be in effect would require him so radically to revise his view of the causes of war and weather as to make nonsense of whatever judgment he might offer; and we are no more able than the agent, to say how he would rank such a gamble among the other propositions in his field of preference.

I take it to be the principal virtue of the present theory, that it makes no use of the notion of a gamble or of any other causal notion. One might suppose that if propositions *A, B, C* are in the preference ranking, then so must be the proposition

(10-3) $$A \text{ if } C,\ B \text{ if not},$$

which asserts that a gamble on *C* is in effect, with possible outcomes *A* and *B;* for one might suppose that the proposition in question would be expressible in our notation as

(10-4) $$AC \vee B\overline{C}\ .$$

But this is not the case: if (10-3) is understood to assert that C causes A and that $\overline{C}$ causes B, then (10-3) and (10-4) are different propositions. To see this, observe that the expected desirability of the gamble (10-3) is

$$\text{(10-5)} \qquad prob\ C\ des\ AC + prob\ \overline{C}\ des\ B\overline{C}$$

while the expected desirability of the proposition (10-4) is

$$\text{(10-6)} \qquad \frac{prob\ AC\ des\ AC + prob\ B\overline{C}\ des\ B\overline{C}}{prob\ AC + prob\ B\overline{C}} .$$

Now I take it that if two propositions are in fact one and the same, they must have the same desirability; but the values of (10-5) and (10-6) may well be different, and therefore I take (10-3) and (10-4) to describe propositions that may well be different.

10.5 Further Comparison with Ramsey's Theory

To avoid confusion that may result from alternative interpretations of the word "if," let us introduce a special symbol

$$\text{(10-7)} \qquad [\ ,\ ,\ ]$$

for the operation which, applied to three propositions

$$X, Y, Z,$$

yields the proposition

$$[X, Y, Z]$$

which asserts that a gamble on Y is in effect, with outcomes X (if Y happens) and Z (if Y fails). Then (10-3) is to be interpreted as $[A, C, B]$.

The symbol (10-7) expresses the key operation of Ramsey's theory. His rule for computing estimated desirabilities of gambles is

$$\text{(10-8)} \qquad des\ [X, Y, Z] = prob\ Y\ des\ XY + prob\ \overline{Y}\ des\ \overline{Y}Z .$$

The condition that Y be ethically neutral relative to X and to Z is simply that both XY and $X\overline{Y}$ be ranked with X and that both ZY and $Z\overline{Y}$ be ranked with Z. Also, if Y is ethically neutral relative to X and to Z, and X and Z are not ranked together, Ramsey's condition for Y to have probability 1/2 is that the gambles $[X, Y, Z]$ and $[Z, Y, X]$ be ranked together, for then we have

$$des\ [X, Y, Z] = prob\ Y\ des\ X + prob\ \overline{Y}\ des\ Z$$

$$des\ [Z, Y, X] = prob\ Y\ des\ Z + prob\ \overline{Y}\ des\ X$$

or, equating desirabilities of gambles that are ranked together and simplifying,

$$prob\ Y = prob\ \overline{Y}$$

Even on Ramsey's generous acceptation, certain gambles must be accounted impossible: thus, we must set

$$[\overline{Y}, Y, Y] = F$$

for there is no gamble on Y in which you get $\overline{Y}$ if you win (= if Y is true) and Y if you lose (= if Y is false). Putting the matter differently, it will always be the case that

(10-9) $$[X, Y, Z] \text{ implies } XY \vee \overline{Y}Z\ .$$

Then, in particular, $[\overline{Y}, Y, Y]$ implies $\overline{Y}Y \vee \overline{Y}Y$, which is F; and any proposition that implies the impossible proposition is itself impossible. Certain difficulties also arise about gambles like $[X, Y, Y]$, which it is impossible to lose. These difficulties can be resolved, and one might develop an interesting version of Ramsey's theory on such a basis. However, our present purpose is merely to point up the contrast between the present theory and Ramsey's by indicating very generally how Ramsey's theory might be put in a somewhat different form: a form in which the three-place operation (10-7) is made to do the work of our operations

(10-10) conjunction, disjunction, denial .

In these terms, the primary difference between Ramsey's theory and ours is the difference between the *causal* operation (10-7) and the *logical* operations (10-10). Ramsey measures his agent's desirability and probability assignments by presenting him with a bewildering variety of possible causal connections between propositions and consequences, some of which are wildly at variance with the agent's notions of how things happen in this world. We perform the corresponding measurements by presenting our agent with a less bewildering variety of entities: with all possible combinations that can be formed by applying the logical operations *not, and,* and *and/or* to any propositions between which he has preferences or between which he is indifferent. It may well be that someone knows that he prefers A to B but does not know whether he prefers $\overline{A}$ to $\overline{B}$, or whether he prefers AB to $A \vee B$; and to that extent our assumptions may represent an idealization, and our theory may be difficult to apply. But it may be argued for our idealization that it would be a gain in clarity and self-knowledge for the agent to determine his preferences among such compounds of A and B. Not so for Ramsey's causal operation: to ask the agent to locate the gamble $[X, Y, Z]$ in his preference ranking when X, Y, and Z are the propositions that (X) there will be a thermonuclear war next week, (Y) this coin will land head up when I toss it, and (Z) there will be fine weather next week, is not to invite him to take pains in the interest of clarity and self-knowledge. To the extent that he can bring himself to consider the gamble seriously, he must entertain alarming and bizarre hypotheses about the person who is offering the gamble: hypotheses that he can only entertain by altering his sober judgments about the causes of war and weather, and thereby altering the very probability assignments which the method purports to measure.

The fact that in the gamble $[X, Y, Z]$ it is the happening of Y or $\overline{Y}$ that is taken to cause the happening of X or of Z is shown by the fact that in computing the desirability of the gamble by (10-8) the weights assigned to the consequences

$$YX, \overline{Y}Z$$

are

$$prob\ Y, prob\ \overline{Y},$$

where the probability of Y is assigned the same value it would have if the gamble were not in effect. Then with X, Y, Z being the propositions about war, the coin, and the weather that were considered in the preceding paragraph, the agent who considers how he ranks the gamble $[X, Y, Z]$ is assigning to the proposition that there will be thermonuclear war next week the probability that he normally assigns to the proposition that the coin will land head up: .500 instead of (say) .001. But the agent who considers how he ranks the corresponding noncasual proposition

$$XY \vee \overline{Y}Z$$

makes no such adjustment: he makes use of his existing beliefs in the independence of war and weather from the outcome of the toss to conclude that

$$prob\ XY = 1/2\ prob\ X$$

and that

$$prob\ \overline{Y}Z = 1/2\ prob\ Z .$$

He judges that

$$des\ XY = des\ X$$

$$des\ \overline{Y}Z = des\ Z$$

since the outcome of the toss will not mitigate the horror of war or enhance his pleasure in the fine weather; and accordingly he substitutes into the general expression

$$des\ (XY \vee \overline{Y}Z) = \frac{prob\ XY\ des\ XY + prob\ \overline{Y}Z\ des\ \overline{Y}Z}{prob\ XY + prob\ \overline{Y}Z}$$

to obtain

$$des\ (XY \vee \overline{Y}Z) = \frac{prob\ X\ des\ X + prob\ Z\ des\ Z}{prob\ X + prob\ Z}\ .$$

If, in particular, he believes that war has probability .001 and desirability $-1{,}000{,}000$, and that fine weather has probability .500 and desirability 1, he arrives at $(-1{,}000 + .5)/.501$ or about $-1{,}995$ as the desirability that he attributes to the proposition $X\overline{Y} \vee \overline{Y}Z$. Throughout this procedure he uses only his actual probability and desirability judgments about the world as he knows it; he has not been required to imagine how his beliefs and desires would be altered if various new causal relationships were introduced.

10.6 Justifying Quantization

I have been at pains to compare the present theory with Ramsey's because Ramsey's theory seems to yield more definite measurements of probability and desirability: once the arbitrary assignments $des\ T = 0$ and $des\ G = 1$ have been made, Ramsey's theory determines all probabilities and desirabilities uniquely, where ours only determines *prob* and *des* as belonging to an infinite family of pairs of assignments. To fix on a particular member of the family is equivalent to choosing a suitable value for the arbitrary parameter c of equations (8-4). Then it might appear that our theory is a partial account of matters concerning which Ramsey's theory gives a complete account; but this appearance is belied by the arguments we have offered against the notion that an agent can usefully judge what his preferences would be if causal connections were altered as they must be in order for certain of Ramsey's gambles to be considered real possibilities. I suggest that the weakness of the present theory is as it should be: that if the boundedness condition (8-18) is satisfied, then there is an inherent indeterminacy in the agent's probability and desirability judgments, corresponding to the latitude we have in choosing the parameter c in equations (8-4); and that it is only in such special circumstances as that in which the unboundedness condition (8-17) is satisfied that this indeterminacy disappears, and the parameter c is forced to assume the value 0. The situation seems strange, but in fact has familiar analogues in situations of the sort that Ramsey considers.

Example 10

If we can ignore the variability of the marginal desirability of money, we might define the subjective probability of a proposition as one hundredth of the largest number of cents that the agent would be willing to pay in order to receive one dollar in case the proposition is true. But then the probabilities of propositions are restricted to the 101 values

$$.00, .01, .02, \ldots, .99, 1.00$$

and by this definition (for a normal agent) the proposition that a certain penny will land head up on the first ten tosses has the same probability, 0, as the proposition F.

Example 11

The definition of example 10 might be weakened and thereby improved, as follows. Suppose that the subjective probability of a proposition is p, and that $100p$ is a whole number. Then the value of p is determined within an interval of length .001 by the following definition. If n is no greater than $100p$ and only then, the agent is willing to pay n cents in order to receive one dollar if the proposition is true. This weakened definition would show that for a normal agent, the subjective probability of the proposition that the first ten tosses of a certain penny will all yield heads has a value in the interval from .00 to .01; it does not lead to the undesirable conclusion that the probability of the proposition in question is 0.

The definition given in example 11 yields a partial determination of subjective probabilities. The determination would be sharper if smaller units of currency than the penny were available, or if we were to countenance bets for more than one dollar. However, as matters stand, the definition has the same sort of indeterminacy that is encountered in chapter 6 when the boundedness condition is satisfied. The analogy is loose, but may nevertheless be helpful in making the indeterminacy that may arise in our system of measurement seem less strange. Of course, that indeterminacy vanishes if the unboundedness condition (8-17) is satisfied; but I know of no reason to suppose that every reasonable man's preference ranking ought to satisfy that condition, unless it be that only then are his degrees of belief determined quite precisely by his preferences.

10.7 Notes and References

It is plausible to regard "Bayesian" desirability theory as originating with Bernoulli: *Specimen theoriae novae de mensura sortis* (1738); see section 1.8 above.

As was noted in section 3.9, the theory presented as Ramsey's in chapter 3 and contrasted in sections 10.4 and 10.5 with the one floated in this book is improved by grafts from Davidson and Suppes's later theory (1957). But those grafts have been further interfered with, in consequence of my identification of the Davidson-Suppes options with propositions. Thus, the features of my account of Ramsey that are criticized in Joseph Sneed's "Strategy and the Logic of Decision" (*Synthese* 16 [1966]: 270–83) depend on that final interference of mine. For an extended treatment of the Ramsey theory, see John M. Vickers's doctoral dissertation (Stanford University, Department of Philosophy, 1962).

11

Probability Kinematics

So far, we have considered probability and desirability statically: we have used the functions *prob* and *des* to describe the agent's attitudes either at a single moment, or during a period within which they do not change. But an adequate account must include changes in the agent's probability and desirability assignments.

11.1 Conditionalization and Its Limits

The beginnings of such an account have already been given in section 5.10, problem 13. There we considered the case in which the agent becomes certain of the truth of a proposition B in which he had formerly had some positive degree of belief short of 1. If the original desirability and probability of a proposition A were

$$des\ A,\ prob\ A\ ,$$

what should the new desirability and probability be after the agent changes his degree of belief in B from $prob\ B$ to 1? The assumption (5-7) was that the new desirability of A should be identical with the old desirability of AB; and it was indicated how, from that assumption, one could deduce that the new probability of A ought to be the ratio of the old probability of AB to the old probability of B. Then the new desirability and probability of A should be

$$des\ AB \qquad \frac{prob\ AB}{prob\ B}$$

The second of these numbers is called the *conditional probability* of A on the evidence B, and will be written either as

$$prob\ (A/B)$$

or as

$$prob_B\ A$$

The process of *conditionalization*—of the agent's changing his subjective probability assignment from *prob* to $prob_E$ upon learning that the evidence-proposition E is true—is thus justified when it is applicable. However, there are cases in which a change in the probability assignment is clearly called for, but where the device of conditionalization cannot be applied because the change is not occasioned simply by learning of the truth of some proposition E. In particular, the change might be occasioned by an observation, but there might be no proposition E in the agent's preference ranking of which it can correctly be said that what the agent learned from his observation is that E is true.

Example 1: Observation by candlelight

The agent inspects a piece of cloth by candlelight, and gets the impression that it is green, although he concedes that it might be blue or even (but very improbably) violet. If G, B, and V are the propositions that the cloth is green, blue, and violet, respectively, then the outcome of the observation might be that, whereas originally his degrees of belief in G, B, and V were .30, .30, and .40, his degrees of belief in those same propositions after the observation are .70, .25, and .05. If there were a proposition E in his preference ranking which described the precise quality of his visual experience in looking at the cloth, one would say that what the agent learned from the observation was that E is true. If his original subjective probability assignment was *prob*, his new assignment should then be $prob_E$, and we would have

$$prob\ G = .30 \qquad prob\ B = .30 \qquad prob\ V = .40$$

representing his opinions about the color of the cloth before the observation, but would have

$$prob\ (G/E) = .70 \qquad prob\ (B/E) = .25 \qquad prob\ (V/E) = .05$$

representing his opinions about the color of the cloth after the observation. But there need be no such proposition E in his preference ranking; nor need any such proposition be expressible in the English language. Thus, the description "The cloth looked green or possibly blue or conceivably violet," would be too vague to convey the precise quality of the experience. Certainly, it would be too vague to support such precise conditional probability ascriptions as those noted above. It seems that the best we can do is to describe, not the quality of the visual experience itself, but rather its effects on the observer, by saying, "After the observation, the agent's degrees of belief in *G, B,* and *V* were .70, .25, and .05."

11.2 The Problem

It is easy enough to cite examples like 1 for the other senses. Transcribing a lecture in a noisy auditorium, the agent might think he had heard the word "red," but still think it possible that the word was actually "led." He might be unsure about whether the meat he is tasting is pork or veal, or about whether the cheese he is smelling is Camembert or Brie. In all such cases there is some definite quality of his sensuous experience which leads the agent to have various degrees of belief in the various relevant propositions; but there is no reason to suppose that the language he speaks provides the means for him to describe that experience in the relevant respects. Or to put the matter in physical terms: in example 1 when the agent looks at the piece of cloth by candlelight there is a particular complex pattern of physical stimulation of his retina, on the basis of which his beliefs about the possible colors of the cloth change in the indicated ways. However, the pattern of stimulation need not be describable in the language he speaks; and even if it is, there is every reason to suppose that the agent is quite unaware of what that pattern is and is quite incapable of uttering or identifying a correct description of it. Thus, a complete description of the pattern of stimulation includes a record of the firing times of all the rods and cones in the outer layer of retinal neurons during the period of the observation. Even if the agent is an expert physiologist, he will be unable to produce or recognize a correct record of this sort on the basis of his experience during the observation.

The remark made at the end of section 4.7 remains true: the theory of preference is ours, but not necessarily the agent's. The theory can be true of an agent even if he is quite ignorant of the theory, and in particular, the language in which we express the propositions in his preference ranking need not be a language that the agent understands. Then, in principle, it is conceivable that we might use propositions about firing patterns of the agent's retinal neurons to record the effects of his visual experience.

All this would be relevant if we were using the theory of preference to describe and interpret the behavior of de facto Bayesians: of agents who in fact behave as if they had preference rankings that satisfy the hypotheses of the uniqueness theorem, although they need not speak of themselves as acting in that way, and perhaps do not use the notions of subjective probability and desirability in discussing their actions. One day Bayesian robots may be built; but at present there are no such creatures, and in particular, human beings are not de facto Bayesians. Bayesian decision theory provides a set of norms for human decision making; but it is far from being a true description of our behavior. Similarly, deductive logic provides a set of norms for human deductive reasoning, but can not usefully be reinterpreted as a description of human reasoning (unless we define reasoning so as to exclude everything that violates the rules of deductive logic, in which case the description becomes true but fragmentary and uninteresting). Indeed, it is because logic and decision theory are woefully inadequate as descriptions that they are of interest as norms. "To stay alive, you must keep inhaling and exhaling" is of little interest as a bit of advice precisely because it is so generally and effortlessly followed by those who can. To serve its normative function, the theory of decision making must be used by the agent, who must therefore be able to formulate and understand the relevant propositions. When Bayesian robots are built, they will conform to Bayesian principles in the way in which steam engines conform to the principles of thermodynamics; but human conformity to Bayesian principles is based on conscious use of them.

Lack of an adequate account of the way in which uncertain evidence can be assimilated into one's beliefs makes for philosophical as well as practical difficulties. C. I. Lewis, taking conditionalization to be the only way of assimilating evidence, concludes that there must exist an "expressive" use of language, by means of which the observer can infallibly record the content of his experience:

> If anything is to be probable, then something must be certain. The data which themselves support a genuine probability, must themselves be certainties. We do have such absolute certainties, in the sense data initiating belief and in those passages of experience which later may confirm it. But neither such initial data nor such later verifying passages of experience can be phrased in the language of objective statement—because what can be so phrased is never more than probable. Our sense certainties can only be formulated by the expressive use of language, in which what is signified is a content of experience and what is asserted is the givenness of this content. [C. I. Lewis (1946), p. 186]

Now it may be that no objective statement can be rendered more than probable by experience. This position is less startling than it might appear at first, if we reflect that .999999 is a probability and that it is short of 1. Then the kind of "practical certainty" that leisurely observation of a bit of cloth in sunlight can lend to the proposition G that the cloth is green can be represented by setting the subjective probability of G after the observation equal to .999999; and this is as close to certainty as makes no odds, practically speaking. But it is quite another matter to conclude from this that there must be a proposition E that can be formulated by an expressive use of language, which has the characteristics that *prob* (G/

$E) = .999999$ and that the observer's degree of belief in E after the observation is 1.

To solve the philosophical problem posed by Lewis's argument, and the practical problems illustrated by example 1, it is necessary to give a positive account of how the agent is to assimilate uncertain evidence into his beliefs. It will not do, in example 1, simply to change the degrees of belief in G, B, and V from .30, .30, and .40, to the new values .70, .25, and .05 without making any further changes in the belief function, for we should then have a belief function which violated the probability axioms. Thus, if *prob* V is .40, *prob* $\overline{V}$ must be .60. If we change *prob* V from the value .40 to the new value .05 without making a compensating change in *prob* $\overline{V}$, we shall have *prob* $(V \vee \overline{V}) = .65$, which contradicts the requirement that the probability of the necessary proposition be 1.

Then the problem is this. Given that a passage of experience has led the agent to change his degrees of belief in certain propositions $B_1, B_2, \ldots, B_n$ from their original values,

$$prob\ B_1, prob\ B_2, \ldots, prob\ B_n$$

to new values,

$$PROB\ B_1, PROB\ B_2, \ldots, PROB\ B_n$$

how should these changes be propagated over the rest of the structure of his beliefs? If the original probability measure was *prob*, and the new one is *PROB*, and if A is a proposition in the agent's preference ranking but is not one of the n propositions whose probabilities were directly affected by the passage of experience, how shall *PROB* A be determined?

11.3 Solution for $n = 2$

As a first step toward answering this question, we consider the important special case in which $n = 2$, and where the pair B_1, B_2 has the form $B, \overline{B}$, so that we would be willing to describe the result of the observation simply by saying that it led the agent to change his degree of belief in some one proposition B, from *prob* B to a new value *PROB* B. (It then goes without saying that degree of belief in $\overline{B}$ changes from *prob* $\overline{B} = 1 -$ *prob* B to *PROB* $\overline{B} = 1 -$ *PROB* B.) This does not mean that the only difference between the old and new belief functions lies in the values they assign to the argument B; but it does mean that the values of *PROB* for all arguments ought to be deducible from a knowledge of (a) the values of *prob* for all arguments, (b) the value of *PROB* for the argument B, and the fact that (c) the change from *prob* to *PROB originated* in B in the sense that for every proposition A in the preference ranking we have

$$\text{(11-1)} \qquad \begin{aligned} &\text{(a) } PROB\ (A/B) = prob\ (A/B) \\ &\text{(b) } PROB\ (A/\overline{B}) = prob\ (A/\overline{B})\ . \end{aligned}$$

Then while the observation changed the agent's degree of belief in B and in certain other propositions, it did not change the conditional degrees of belief in any propositions on the evidence B or on the evidence $\overline{B}$.

Example 2: The mudrunner
A racehorse performs exceptionally well on muddy courses. A gambler's degree of belief in the proposition A that the horse will win a certain race should change if a fresh weather forecast leads him to change his degree of belief in the proposition B that the course will be muddy. However, the forecast should have no effect on his degrees of belief in the proposition that the horse will win conditionally on the course being muddy, or on its not being muddy.

Now suppose that neither *prob B* nor *PROB B* has either of the extreme values, 0 and 1; suppose that *prob* and *PROB* satisfy the probability axioms (5-1); and define conditional probability in the usual way,

$$prob\ (A/B) = \frac{prob\ AB}{prob\ B}$$

$$PROB\ (A/B) = \frac{PROB\ AB}{PROB\ B}$$

Then it is straightforward to show that condition (11-1) is equivalent to the condition that for all propositions A in the preference ranking we have

$$\text{(11-2)} \quad PROB\ A = prob\ (A/B)\ PROB\ B + prob\ (A/\overline{B})\ PROB\ \overline{B}.$$

Now (11-2) is a formula of the required sort. It determines all values of *PROB* in terms of knowledge of all values of *prob* and a knowledge of the value that *PROB* assigns to the argument *B;* and it is applicable in exactly the case where the change from *prob* to *PROB* originates in B in the sense that (11-1) holds for all A in the preference ranking.

Example 3: The mudrunner, continued
In example 2, suppose that the gambler's degree of belief in the proposition A that the horse will win conditionally on the proposition B that the track will be muddy is .8, but that his degree of belief in A conditionally on $\overline{B}$ is only .1; and suppose that a fresh weather forecast leads him to change his degree of belief in B from .3 to a new value .6. Then by (11-2) his degree of belief in A should assume the new value

$$(.8)(.6) + (.1)(.4) = .52\ .$$

Notice that by (11-3) (a) below, his original degree of belief in A must have been

$$(.8)(.3) + (.1)(.7) = .31 .$$

11.4 Relevance

The probability axioms, together with the definition of conditional probability, imply that if neither *prob B* nor *PROB B* is 0 or 1 we have

(11-3) (a) $prob\, A = prob\,(A/B)\, prob\, B + prob\,(A/\overline{B})\, prob\, \overline{B}$
(b) $PROB\, A = PROB\,(A/B)\, PROB\, B + PROB\,(A/\overline{B})\, PROB\, \overline{B}$.

Thus, (11-2) is derived from (11-3) (b) in conjunction with (11-1). Now we can get some further insight into the effects of the change from *prob* to *PROB* by subtracting equation (11-3) (a) termwise from equation (11-2) and using the facts that

$$prob\, \overline{B} = 1 - prob\, B \qquad PROB\, \overline{B} = 1 - PROB\, B$$

to obtain

$$PROB\, A - prob\, A = (PROB\, B - prob\, B)[prob\,(A/B) - prob\,(A/\overline{B})]$$

This last equation holds whenever the change from *prob* to *PROB* is governed by (11-2), and expresses the change in the probability of an arbitrary proposition A as the result of multiplying the original change, in the probability of B, by the factor

(11-4) $$rel\,(A/B) = prob\,(A/B) - prob\,(A/\overline{B})$$

which we may call the *relevance to A of B* (relative to *prob*). Then the relationship between the changes in probability of A and of B is

(11-5) $$PROB\, A - prob\, A = (PROB\, B - prob\, B)\, rel\,(A/B) .$$

Example 4: The mudrunner, concluded

In example 3, *rel* (A/B) is $.8 - .1$, so that the increase $.52 - .31 = .21$ in the gambler's degree of belief in A must be seven tenths of the increase $.6 - .3 = .3$ in his degree of belief in the proposition B in which the change originated.

It is evident from equation (11-5) that when the change from *prob* to *PROB* originates in an increase in the agent's degree of belief in B, then the probability of a proposition A increases, decreases, or remains the same accordingly as the relevance of B to A is positive, negative, or zero; and it is evident that the effect on A is opposite if the change originates in a decrease in the agent's degree of belief on B. This is as it should be if positive, negative, and null values of *rel* in (11-4) correspond to what we would ordinarily mean by "positive relevance," "negative relevance," and "irrelevance." It appears that this is so. Certainly, the sign of *rel* (A/B) will always agree with that of the measure of relevance that Carnap has proposed,

$$prob\ AB - prob\ A\ prob\ B\ ,$$

for it is straightforward to verify that we have

$$rel\,(A/B) = \frac{prob\ AB - prob\ A\ prob\ B}{prob\ B\ prob\ \overline{B}}$$

as long as *prob B* is neither 0 nor 1.

11.5 Comparison with Conditionalization

It is plain that when *PROB B* is close to 1, equation (11-2) determines *PROB A* as approximately *prob* (A/B) for each proposition A, and thus determines *PROB* as approximately the same assignment as $prob_B$. Then conditionalization is a limiting case of the present more general method of assimilating uncertain evidence, and the case of conditionalization is approximated more and more closely as the probability of the evidence B, approaches 1. It would make no detectable difference if we adhered to C. I. Lewis's line and refused ever to assign subjective probability 1 to an objective proposition other than T, provided we were willing to assign such propositions values that are "practically" 1, e.g., .999999. And there would be a certain advantage in taking this line, for as long as *PROB B* is neither 0 nor 1, the change from *prob* to *PROB* is reversible in a way in which the change from *prob* to $prob_B$ is not.

To see that the change from *prob* to $prob_B$ is irreversible by conditionalization, suppose that the agent uses conditionalization to change from the belief function *prob* to a new belief function *PROB* in response to an observation, the effect of which is to convince him of the truth of B; but suppose that a further observation convinces him that he completely misconstrued the first observation, which in fact yielded no information. Then he would like to change from the belief function *PROB* back to his original belief function *prob*. However, he cannot do this by conditionalization: There is no proposition C which has the characteristic that $PROB_C$ is identical with the original assignment *prob*, except in the trivial case where $prob\ B = 1$, in which case, he was sure of the truth of B even before the first observation. To see that no such C exists, note that $PROB_C\ B$ must equal *prob B* if the reversal is to take place, and that for any A we have $PROB\ A = prob\ (A/B)$. Then we must have $PROB_C B = prob\ (B/BC)$

= 1, so that if $PROB_C B$ is to be the same number as *prob B*, that number must have been 1.

On the other hand the change from *prob* to *PROB* via (11-2) is reversible by another application of (11-2) as long as neither *prob B* nor *PROB B* has either of the extreme values, 0 and 1. In particular, suppose that the agent makes an observation of which the effect is to change his degree of belief in *B* from *prob B* to a new value *p*. Then by (11-2), his degree of belief in an arbitrary proposition *A* will change from *prob A* to

$$PROB\ A = prob\ (A/B)\ p + prob\ (A/\overline{B})(1 - p)\ .$$

Suppose that a further observation has as its effect a change in his degree of belief in *B* from *p* back to *prob B*. Applying (11-2) again, his degree of belief in an arbitrary proposition *A* will become

$$PROB\ (A/B)\ prob\ B + PROB\ (A/\overline{B})\ prob\ \overline{B}$$

or, by (11-1) and (11-3) (a),

$$prob\ A\ .$$

Then the method of changing belief that is given by (11-2) has the advantage of reversibility: mistakes can be erased.

11.6 Solution for Finite *n*

Now what of the general case where the effect of the observation is to simultaneously change the agent's degrees of belief in two or more propositions? In the case of conditionalization, there is no separate difficulty here, because the result of conditionalizing relative to *B* is to change *prob* (. . .) to *prob* (. . ./*B*), and the result of then conditionalizing relative to *C* is to change *prob* (. . ./*B*) to *prob* (. . ./*BC*); and this is exactly what the result would have been if the agent had straightway conditionalized relative to *BC*. Then the effect of first changing the probability of B to 1, and then changing the probability of *C* to 1 can be achieved by immediately changing the probability of a single proposition *BC* to 1. It is not so for the more general technique of assimilating changes in belief via (11-2). There the effect of first changing the probability of *B* to *p* and then changing the probability of *C* to *q* cannot generally be produced by changing the probability of some one proposition to some new value *r*. Nor need it always be the case that the result of two applications of (11-2) is independent of the order: the belief function obtained from *prob* by first changing the probability of *C* to *p* and then changing the probability of *C* to *q* need not be the same as the belief function that would have been obtained if first the probability of *C* had been changed to *q*, and then the probability of *B* had been changed to *p*. But in the case of conditionalization, order is irrelevant: $prob_{BC}$ is always the same assignment as $prob_{CB}$ because BC is always the same proposition as CB.

In generalizing (11-2) to cover the general case in which the change in belief originates in a set of n propositions

(11-6) $$B_1, B_2, \ldots, B_n$$

it will be necessary to consider not the B's themselves, but rather the collection of all conjunctions of form

$$C_1C_2 \ldots C_n$$

where each of the C's is identical either with the corresponding B or with its denial, and where the whole conjunction is not identical with the impossible proposition F. Each such conjunction will be called an *atom* of the set (11-6). There may be as many as 2^n distinct atoms of such a set.

Example 5: Observation by candlelight, continued

In example 1, n is 3 and B_1, B_2, B_3 are the propositions G, B, V. There are $2^3 = 8$ expressions of form $C_1 C_2 C_3$:

$$GBV,\ GB\overline{V},\ G\overline{B}V,\ G\overline{B}\,\overline{V},\ \overline{G}BV,\ \overline{G}B\overline{V},\ \overline{G}\,\overline{B}V,\ \overline{G}\,\overline{B}\,\overline{V}$$

However, if we interpret G as meaning that the cloth is green *all over*, and, similarly, for B and V, and if we suppose that for some reason it is impossible that the cloth have any color other than these three, the eight expressions denote (with repetitions) only four distinct propositions,

$$F,\ F,\ F,\ G,\ F,\ B,\ V,\ F$$

Thus, the first of the eight conjunctions asserts that the cloth is at once green, blue, and violet all over, and is therefore impossible; the fourth of them asserts that the cloth is green all over and then redundantly denies that it is blue all over and that it is violet all over; and the last of them is impossible because we have assumed that $G \vee B \vee V$ is necessary. Then the atoms of the set G, B, V are those three propositions themselves.

11.7 Origination, Closure

Suppose that the distinct atoms of the set (11-6) are the propositions

$$A_1, A_2, \ldots, A_m$$

Then the appropriate generalization of the formula (11-2) will prove to be

(11-7) $$PROB\ A = prob\ (A/A_1)\ PROB\ A_1 + prob\ (A/A_2)\ PROB\ A_2 + \ldots + prob\ (A/A_m)\ PROB\ A_m$$

and the appropriate generalization of formulas (11-1) will prove to be

$$\text{(11-8)} \quad PROB\ (A/A_i) = prob\ (A/A_i) \text{ for each } i = 1, 2, \ldots, m$$

which defines what we shall mean by saying that the change from *prob* to *PROB originated* in the set (11-6). It is straightforward to show that conditions (11-7) and (11-8) are equivalent if both *prob* and *PROB* satisfy the probability axioms and none of the m numbers $prob\ A_i$ is either 0 or 1.

Example 6: Observation by candlelight, concluded

If the observation described in example 1 changes the probabilities of G, B, and V from .30, .30, and .40 to .70, .25, and .05, but does not change the probabilities of any propositions conditionally on G, or on B, or on V, then the change from *prob* to *PROB* does originate in the set G, B, V, and the atoms of that set are those three propositions themselves. Then by (11-7) we have

$$PROB\ A = .70\ prob\ (A/G) + .25\ prob\ (A/B) + .05\ prob\ (A/V)\ .$$

Note that since distinct sets of propositions can have the same set of atoms, there is a certain latitude in the choice of a set (11-6) in which the change from *prob* to *PROB* is viewed as originating.

Example 7: Observation by candlelight, reviewed

The change described in example 6 as originating in the set G, B, V might equally well be regarded as originating in the set

$$B_1 = G \vee B \quad B_2 = G$$

for the four conjunctions of form C_1C_2 would then be

$$(G \vee B)G = G \quad (G \vee B)\overline{G} = B \quad (\overline{G \vee B})G = F \quad (\overline{G \vee B})\overline{G} = V$$

so that we have the same set of atoms as before, and formula (11-7) will yield the same function *PROB* as before.

Let us speak of a set of propositions as being *closed* if it is closed under the operations of conjunction, disjunction, and denial, in the sense that whenever A and B are members of the set, so are

$$\overline{A}, \overline{B}, AB, A \vee B\ .$$

Thus, the set consisting of F together with all of the propositions in the agent's preference rankink is a closed set. And let us speak of the *closure* of a set $\mathcal{B}$ of propositions as being the least inclusive closed set to which every member of $\mathcal{B}$ belongs. If $\mathcal{B}$ is finite, its closure will consist of F, together with the atoms of $\mathcal{B}$, together with all disjunctions of two or more atoms of $\mathcal{B}$. Then if $\mathcal{B}$ has m atoms, the closure of $\mathcal{B}$ will consist of 2^m propositions.

If the members of $\mathcal{B}$ are $G \vee B$ and G, as in example 7, and if $V = \overline{G \vee B}$, then the atoms of $\mathcal{B}$ are G, V, and B, and its closure will consist of the eight propositions

$$F,\ G,\ B,\ V,\ G \vee B,\ G \vee V,\ B \vee V,\ T\ .$$

The simplest cases of change from *prob* to *PROB* are those for which we can find a proposition B such that (11-1) holds for every proposition A in the agent's preference ranking. The change is then said to originate in the proposition B or, alternatively, in the set consisting of the two propositions B, $\overline{B}$. This notion of origination was then generalized to cover cases where the change is said to originate in a finite set (11-6) of propositions: these are the cases where (11-8) holds for every proposition A in the agent's preference ranking. It only remains to generalize the notion of origination to cover cases where the change would properly be said to originate in an infinite set $\mathcal{B}$.

11.8 The Continuous Case

Example 8: Observation by candlelight, generalized

The possible hues of a piece of cloth of uniform hue form a continuum which can be described conveniently in terms of the wavelength of reflected sunlight, measured in millimicrons (mμ). The range from green to violet falls within the interval from 550 to 350 on this scale, and it is conceivable that the observer thinks of hues in terms of wavelengths instead of (or, in addition to) thinking of them in terms of color words in English. His beliefs about the hue of the cloth before and after the observation might then be represented by two probability distributions as illustrated in figure 11.1, where B_x is the proposition that the average wavelength of the light that would be reflected from the cloth in sunlight is between 350 mμ and x mμ. Presumably, the probabilities of all propositions of form $B_y\overline{B_x}$ where y is greater than x approach 0 as y approaches x: such propositions say that the wavelength is between x and y mμ, and as the interval shrinks to 0, so does the probability that the average wavelength is in that precise interval. In particular, for any x, the probability of the proposition that the wavelength is exactly x is 0, both before and after the observation. In terms of wavelengths, the propositions V, B, and G of example 1 are $B_{425}\overline{B_{375}}$, $B_{475}\overline{B_{425}}$, and $B_{525}\overline{B_{475}}$; according to figure 11.1 (a), the probabilities of these propositions before the observation are .40, .30, and .30; and according to figure 11.1 (b), the probabilities of the same propositions after the observation are .05, .25, and .70. Thus, *prob B* is *prob* $B_{475}\overline{B_{425}}$ or *prob* $\overline{B_{475}}$ − *prob* B_{425}, and by figure

Fig. 11.1

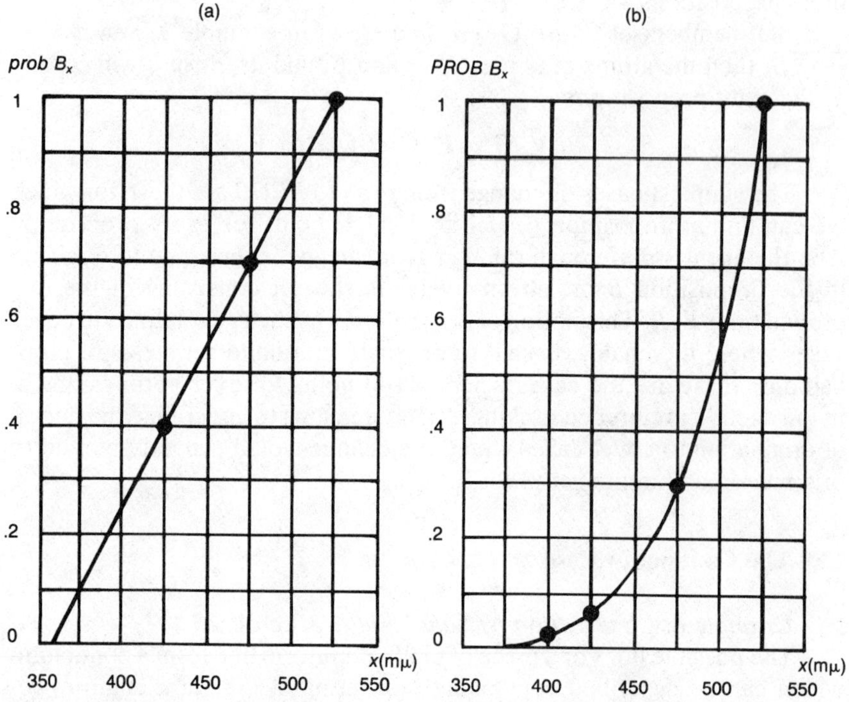

11.1 (a) we have *prob* B_{475} = .70 and *prob* B_{425} = .40, so that indeed *prob* B = .30.

In cases of the sort considered in example 8, the set $\mathcal{B}$ in which the change is taken to originate would be infinite; thus, in Example 8, $\mathcal{B}$ might be taken to be the set of all propositions of form B_x where x lies between 350 and 550. Formula (11-7) can be applied approximately, in such cases.

$$A_1 = B_{375}\overline{B_{350}} \quad A_2 = B_{400}\overline{B_{375}} \quad \ldots \quad A_8 = B_{550}\overline{B_{525}}$$

These eight A's behave like atoms in that they are incompatible and together exhaust the range within which x must lie; and they satisfy conditions (11-8) approximately, in the sense that for any proposition A in the preference ranking the numbers *prob* (A/A_i) and *PROB* (A/A_i) will be approximately the same, for each i = 1, 2, . . . , 8. (The approximation would be even better if we were to divide the interval into sixteen equal subintervals, obtaining sixteen such A's.) Now equation (11-7) can be used to get approximate values of *PROB A* for any proposition A in the preference ranking.

Example 9

Divide the interval from 350 to 550 into eight equal subintervals, and define eight propositions which behave approximately like atoms

$$\begin{aligned} PROB\ A &\approx prob\ (A/A_1)\ PROB\ A_1 + prob\ (A/A_2)\ PROB\ A_2 \\ &\quad + \ldots + prob\ (A/A_8)\ PROB\ A_8 \\ &= prob\ (A/A_1)(0) + prob\ (A/A_2)(.01) + \ldots + prob\ (A/A_8)(0) \end{aligned}$$

To obtain more accurate values of *PROB A*, use sixteen (or 32, or 64) pseudoatoms.

In such cases as these, we may think of the agent's change in belief as originating in the closure of the set $\mathcal{B}$, and as being propagated over the remainder of the structure of his beliefs by the method illustrated in example 9. To discuss the matter more rigorously and generally, it is necessary to use the notion of integration over abstract spaces; but the essential ideas are illustrated in example 9, since the integrals in question would be defined by exactly the sort of process of successive approximation that was indicated there.

11.9 Probabilistic Acts; Trying

The situation that we have been studying in relation to probabilistic observations has its parallel in the case of probabilistic acts. It may be that the agent decides to perform an act which is not simply describable as *making the proposition B true*, but must be described as changing the probabilities of two or more propositions (11-6) from

$$prob\ B_1,\ prob\ B_2,\ \ldots,\ prob\ B_n$$

to a new set of values,

$$PROB\ B_1,\ PROB\ B_2,\ \ldots,\ PROB\ B_n\ .$$

In the simplest cases, where $n = 2$, where B_1 is some good proposition B, and where B_2 is the bad proposition $\overline{B}$, we speak of the agent as *trying to make B true;* and where *PROB B*, the probability that *B* would have if the agent decided to perform the act, is very close to 1, we may speak of the agent as believing it to be in his power to make *B* happen if he chooses. Here *PROB* reflects the agent's belief, before seeing the act's outcome, in his ability to make *B* true by the act in question. (Then the change in the agent's degree of belief from *prob B* to *PROB B* is what Anscombe calls "knowledge without observation.") But where we and the agent are willing to speak of him as trying to make *B* true, there need be no corresponding proposition in the agent's preference ranking, just as in example 1 there need be no such proposition as the cloth's looking green. There, the situation that we and the agent might describe as the cloth's looking green is represented as a change in the agent's degree of belief in the proposition G that the cloth *is* green from .30 to .70, rather than by a change in the agent's degree of belief in some corresponding expressive proposition E (that the cloth looks green) from some value *prob* E to 1. We do not assume that propositions like E occur in the agent's preference ranking. Here the "trying" idiom has the same sort of function in connection with acting that the "looking" idiom has in connection with observing. We do not suppose that where we and the agent are willing to speak of him as trying to make B true, his preference ranking must contain a proposition

$$E = \text{that the agent tries to make } B \text{ true .}$$

Rather, to speak of the agent's trying to make B true is to speak of his performing an act of which he takes the net effect to be an increase in the probability of B. We or the agent may speak in this way without thereby assuming the existence of a proposition E in his preference ranking for which we have

$$(11\text{-}9) \qquad PROB\ E = 1 \qquad PROB\ B = prob\ (B/E)$$

where *PROB* is the agent's belief function after he decides to perform the act. Thus, trying to hit the bullseye may be an act without there being any proposition that plays E to hitting the bullseye's B, above.

Example 10: The comforter

The agent is trying to comfort a lady whose cat has been killed. This may consist in any of a variety of acts, such as giving her another cat, holding her hand, or saying "He was getting old and stiff, anyway." And there are many ways of performing the last-mentioned act, of saying the words, some of which would be more likely to produce comfort than others: variations in volume of voice, proximity of speaker to hearer, and facial expression might all be important. The agent might have an accurate sense of how he is speaking the comforting words in each of these respects without being able to verbalize it. Thus, he might be tacitly aware that saying the words in a loud voice from across a room with his facial muscles relaxed might have a disturbing effect; and he might be able to control his distance, volume, and facial expression in the relevant ways; and he might nevertheless be unable to produce or recognize a true description of what he is doing, in terms of distance, volume, and facial expression. The overall effect of his tacit awareness of what he is doing might well be that he takes his act of speaking the comforting words to impart a certain high probability *PROB B* to the woman's being comforted, without his being able to characterize the act by means of a proposition E in his preference ranking for which (11-9) holds. (The importance of such tacit knowledge in epistemology—knowledge *how,* unaccompanied by knowledge *that*—has been stressed by Michael Polanyi.)

11.10 Observation; Meaning

It has been held by many empiricists that the propositions expressible in the agent's language can usefully be classified as *observational* or not, and that the nonobservational propositions derive their significance from their logical relationships with the observational ones, in virtue of which nonobservational propositions can be confirmed or disconfirmed to various degrees on the basis of the agent's observations. (Something like this position was maintained by Carnap, e.g., in "Testability and Meaning," *Philosophy of Science* 3 [1936]: 419–71 and 4 [1937]: 1–40.) In view of what we have said here it seems unpromising to construe the observational propositions as the ones whose truth or falsehood can be ascertained with certainty by making suitable observations; but the notion might still be defined in terms of changes that originate in sets (11-6) of propositions in the sense that (11-8) holds for all propositions A that are expressible in the language. The *observational basis* of a certain language for a certain agent might then be taken to be the smallest set of propositions expressible in the language which has the characteristic that any possible change in the agent's belief function which is occasioned by an observation can be described as originating in some set (11-6) of propositions, all of which belong to the observational basis.

Similarly, there will be an *actual basis* for a given agent and a given language, consisting in the smallest set of propositions expressible in the language which has the characteristic that the effect of any possible act on the agent's belief function can be described as a change which originates in a set (11-6) of propositions, all of which belong to the actual basis. It is to be expected that the actual basis will be a proper subset of the observational basis, which will be in turn a proper subset of the collection of all propositions that are expressible in the language. Roughly speaking, the actual basis is the set of propositions whose truth values the agent thinks he can directly influence by his acts, and the observational basis is the set of propositions whose truth values the agent thinks capable of having a direct influence on his belief function via observations.

No doubt, paradigm cases play an important role in discussions of meaning. To know the proper use of the English color words "green," "blue," and "violet" is in no small part to be able to apply them correctly to objects observed in sunlight whose colors fall well within the ranges of hues that are covered by the three words; and in such cases it would be grotesque to balk at describing the agent as having seen that the objects were green, blue, and violet, and as having seen that the corresponding propositions are true. But often enough the agent will have to deal with unparadigmatic cases, in which (say) the light conditions or the actual hues are such as to make it clearly inappropriate for him or us to describe his observation in such simple terms. Here it becomes necessary to recognize the components of meaning that must be described in terms of degrees of belief short of certainty. Thus, suppose that the light is dim and that the hue of the object being observed is between violet and blue: the wavelength of reflected midday sunlight would be 425 mμ, let us say. Then doubt might be cast on the agent's mastery of the English color words if he responded to the observation by changing the probabilities of the propositions V, B, and G from .4, .4, and .2, to .2, .2, and .6. It seems fair to put the matter in this way: mastery of the English color words involves a working knowledge—tacit or otherwise—of the fact that blue comes between violet and greet on the color spectrum. And more generally, mastery of a language involves an accurate sense of the *confirmational proximity* of propositions, in the light of which certain changes in belief would be viewed as inappropriate to the observation in question.

It is true that such a sense of confirmational proximity would be acquired through experience, but this need not count against its status as a condition for mastery of the language, and as an ingredient in a grasp of the meanings of the relevant words. In "logical reconstruction" of language as in actual construction of a robot, one might draw a sharp line between aspects of the agent's belief function that are built into the semantics of his language and those which arise in response to experience. But by and large, we speak unreconstructed languages and are never done with the business of learning them; and it is some part of the business of epistemology to treat of human beings as they are.

11.11 Notes and References

The passage from C. I. Lewis in section 11.2 appears in *An Analysis of Knowledge and Valuation,* La Salle, Illinois: Open Court, 1946) p. 186. There was more of the same in Lewis's *Mind and the World-Order* (1929; reprint ed. New York: Dover, 1956), chapter 10 (e.g., pp. 328–29). A three-way discussion of the matter, "The Experiential Element in Knowledge," by Hans Reichenbach, Nelson Goodman, and C. I. Lewis, appears in the *Philosophical Review* 61 (1952): 147–75. Appeals to Lewis's "expressive" use of language, or to special forms of speech in which private contents of experience can infallibly be described, have been less common in philosophical literature since the critical work of Ludwig Wittgenstein,

Philosophical Investigations (Oxford:Basil Blackwell, 1953), see par. 256ff., and J. L. Austin, *Sense and Sensibilia* (Oxford: Oxford University Press, 1962), chapter 10.

Carnap discusses the measure of relevance that is mentioned at the end of section 11.4 in chapter 6 of *The Logical Foundations of Probability* (Chicago: University of Chicago Press, 1950; second edition, 1962).

G. E. M. Anscombe discusses knowledge without observation (see section 11.9) in *Intention* (1st ed. Oxford: Basil Blackwell, 1957; 2d ed. Ithaca, N.Y.; Cornell University Press, 1963). For Polanyi on tacit knowledge, see *Personal Knowledge* (Chicago: University of Chicago Press, 1958).

This chapter stems from chapter 3 ("Elementary Dynamics of Belief") of my Ph.D. dissertation, "Contributions to the Theory of Inductive Probability" (Princeton University, 1957). For two other treatments of this material, see my papers "Probable Knowledge," in *The Problem of Inductive Logic,* ed. Imre Lakatos (Amsterdam: North-Holland, 1968), reprinted in the Kyburg and Smokler collection cited at the beginning of section 3.9, and "Dracula Meets Wolfman: Acceptance vs. Partial Belief," in *Induction, Acceptance, and Rational Belief,* ed. Marshall Swain (Dordrecht: D. Reidel, 1970).

Isaac Levi has been an early and persistent critic of my suggestion that probability kinematics provides a solution to the problem posed in section 9.2. For the opening blast, see his "Probability Kinematics," *British Journal for the Philosophy of Science* 18 (1967): 197–209, to which Henry E. Kyburg, Jr., and William Harper reply in *BJPS* 19 (1968): 247–58. For a later volley, see Levi's contribution (in his aspect as Wolfman) to the Swain collection cited above. Levi's own view is that we ought to assign probability 1 to background hypotheses whose credentials are not currently suspect, while remaining ready to scrutinize such credentials should they be seriously questioned. For my part, I see no obstacle to reasoning from hypotheses one does not fully believe, or even from hypotheses that are under serious suspicion—as in refutations by reductio ad absurdum.

For a different sort of challenge to the kinematical scheme of sections 11.3–11.6, see Hartry Field, "A Note on Jeffrey Conditionalization," *Philosophy of Science* 45 (1978): 361–67, and the reply by Daniel Garber, "Field and Jeffrey Conditionalization," *Philosophy of Science* 47 (1980): 142–45.

For two related attempts to establish the credentials of the kinematical scheme, see Paul Teller, "Conditionalization and Observation," *Synthese* 26 (1973): 218–58, and Brad Armendt, "Is There a Dutch Book Argument for Probability Kinematics?" *Philosophy of Science* 47 (1980): 583–88.

Lately, much attention has been given to the connection between the present kinematics and E. T. Jaynes's scheme of maximum entropy inference, which he has proposed in a series of publications since 1957, e.g. in "Foundations of Probability Theory and Statistical Mechanics," *Delaware Seminar in the Foundations of Physics,* ed. Mario Bunge (New York: Springer, 1967). The connection was first examined in Sherry May's Ph.D. thesis, "On the Application of a Minimum Change Principle to Probability Kinematics" (University of Waterloo, Faculty of Mathematics, November 1973); see also Sherry May and William Harper, 'Toward an Optimization Procedure for Applying Minimum Change Principles in Probability Kinematics," in W. Harper and C. Hooker, ed., *Foundations of Probability Theory, Statistical Inference, and Statistical Theories of Science,* vol. 1 (Dordrecht: D. Reidel, 1976). The thought is that the right new belief function *PROB* is to be sought among those that satisfy constraints imposed by observation and is to be identified as the one among them that is closest to the old belief function *prob* in some sense of "close." As Persi Diaconis and Sandy Zabell show in section 5 of "Updating Subjective Probability" (*Journal of the American Statistical Association,* 77, no. 4 [1982]), the present kinematical scheme yields the closest new belief function on several common understandings of closeness, among which is the Jaynes-like relative entropy, which measures the distance between *prob* and *PROB* as

$$\Sigma_i \, PROB \, A_i \log \frac{PROB \, A_i}{prob \, A_i}$$

relative to the partitioning of T into incompatible, exhaustive propositions A_i. See P. M. Williams, "Bayesian Conditionalization and the Principle of Minimum Information," *British Journal for the Philosophy of Science* 31 (1980): 131–44, for an account of matters from the relative entropy point of view, and see Bas van Fraassen, "A Problem for Relative Information Minimizers in Probability Kinematics," *British Journal for the Philosophy of Science* 32 (1981): 367–79 for a demurral. The requirement that the A_i be incompatible and exhaustive can be relaxed without disturbing the meaningfulness of the question of distance between *prob* and *PROB,* so that minimum-distance inference is a genuine generalization of inference via (11-7): see Bas van Fraassen, "Rational Belief and Probability Kinematics," *Philosophy of Science* 47 (1980): 165–78, and Diaconis and Zabell. Warning: different senses of "close" can yield nontrivially different generalizations, even though they agree with each other and with (11-7) when the A_i form a partition of T. See Diaconis and Zabell for details and for further references to the literature.

It is straightforward to verify that the present kinematical scheme is not generally commutative: if we apply (11-7) twice, so as to make the probabilities of the A_i be a_i and (then or before) make the probabilities of the B_j be b_j, the resulting probability measure may be one thing or another depending on the order of the two applications. That is as it should be, for just after the A-application, one's degrees of belief in the A_i *are* a_i, no matter what they were before, and after the B-application, the degrees of belief in the B_j *are* b_j, no matter what they were before. Then if the A-application comes second, degrees of belief in the A_i will then be a_i, but if the order is reversed, the B-application may disturb the earlier setting of the probabilities of the A_i at a_i. (This cannot happen if the A_i and B_j are independent relative to the initial probability function, in the sense that for all i and j, *prob* A_iB_j is simply the product of *prob* A_i with *prob* B_j.) For much more about these matters, see section 3 of Diaconis and Zabell's forthcoming article.

The notion that the kinematical scheme *ought* to be generally commutative (Domotor, *Philosophy of Science* 47 [1980]: 395) stems from a conflation of two attitudes toward The Given: the one that motivates the present kinematical scheme, and another, that moves Hartry Field to reparametrize it ("Note on Jeffrey Conditionalization"). The two attitudes are represented in the correspondence with Carnap recorded in my paper "Carnap's empiricism," in *Induction, Probability, and Confirmation,* ed. Grover Maxwell and Robert M. Anderson (Minneapolis: University of Minnesota Press, 1975). I see Carnap's attitude as a doomed attempt to make sense of The Given in probabilistic terms, with the α_i and β_j representing the purely observational "inputs" (stripped of the effects of memory and inductive relevance that are packed into the belief functions *prob* and *PROB* that characterized the agent's epistemic states at the times of the observations). As I read him, Field sees α_i and β_j as constructs out of the a_i and b_j and the functions *prob* and *PROB,* and he has a particular proposal (faulted by Garber, op. cit.) for carrying out the construction. With Field, I see the α_i and β_j as artifacts, but unlike him, and like Garber, I suspect them of being epistemological geegaws that do no work. I prefer to make do with the a_i and b_j, which represent what Field and Domotor see as the probabilistic output of the observer-as-black-box, when the inputs are the α_i and β_j.

12

Induction and Objectification

We shall now examine some aspects of probability kinematics that are the same, whether changes in belief are propagated by the method of conditionalization or by the more general methods of chapter 11. Accordingly we shall restrict attention to the simplest case, that of conditionalization, where the result of an observation is to change the agent's belief function from *prob* to $prob_B$ for some proposition B of which the truth has been ascertained by the observation.

12.1 Belief: Reasons versus Causes

It would be a mistake to speak of the "testimony" of the senses in describing the relationship between the observer's sense experience and the observational proposition B which is involved in the transition from *prob* to $prob_B$. The passage of sense experience to which the observer responds by believing B to degree 1 is conclusive, since B is then believed with certainty; but if it is conclusive evidence, it is no sort of testimony—in particular, it is not testimony of an unshakably reliable character. The experience is not a *reason* for believing B in the way in which truth of X would be, when X implies B or when $prob\ (B/X) = 1$. And while it might be argued that some or all reasons for believing something are also causes of that belief, it is clear that not all causes of belief are reasons for belief: witness, wishful thinking. Nor is it always a defect when beliefs are caused but not reasoned: when an observer justifies his belief that the sun is shining by saying, "I see it," he is citing the cause of his belief. He is not blind; he is outdoors, with his eyes open; and in the circumstances he has no choice but to believe that the sun is shining, for he sees that it is. (Similarly, if he were standing on a chair which collapsed he would have no choice but to fall, and the collapse of the chair would have caused his fall but would not have provided a reason for it.) If the observer wishes to convince a doubter that the sun is indeed shining, he can do no better than to get the doubter to come outdoors and see for himself: to put himself into a position where his degree of belief in B will be forced to 1 by his sense experience. The doubter might also be convinced or nearly convinced by reasons, for example, by the consideration that it is noon in Fresno, California, on July 15, and that there, the sun is nearly always out then; but in such cases reasons are inferior to the right sorts of causes as proper sources of beliefs.

We shall now consider cases in which the agent's belief function changes from *prob* to $prob_B$ as the result of an observation, where the agent's conclusive belief in B is caused by the observation, is unreasoned, and is justified by the consideration that the observation is of the paradigmatic sort to which any normal speaker of the language in which B is expressed would respond by believing B, willy-nilly. If A is a proposition in the agent's probability field whose probability changes from a low or middling value *prob* A to a high value $prob_B\ A$, as a result of the observation, the agent may be said to believe A as a result (an indirect result) of the observation, and B will be functioning as evidence for A, so that the agent's unreasoned belief in B is his reason for believing A.

12.2 Bayes's Theorem

An important class of cases is that in which there are a number of hypotheses

$$A_1, A_2, \ldots, A_m$$

which are mutually exclusive and collectively exhaustive in the sense that we have

$$prob\ A_i A_j = 0 \text{ if } i \neq j\ , \tag{12-1}$$

$$prob\ (A_1 \vee A_2 \vee \ldots \vee A_m) = 1\ ; \tag{12-2}$$

where the proposition B has various probabilities

$$prob\ (B/A_1),\ prob\ (B/A_2),\ \ldots,\ prob\ (B/A_m) \tag{12-3}$$

conditionally on the various hypotheses; where there is wide intersubjective agreement about (12-1), (12-2), and the numbers (12-3) in the sense that in these respects the agent's belief function agrees with those of the other agents with whom he is concerned; but where before the observation there is little intersubjective agreement about the unconditional probabilities of the hypotheses, so that the numbers

$$prob\ A_1,\ prob\ A_2,\ \ldots,\ prob\ A_m \tag{12-4}$$

which represent the agent's degrees of belief in the hypotheses before the observation may differ markedly from the corresponding numbers for other agents. In such cases the evidence B may have the effect of promoting intersubjective agreement about the probabilities of the hypotheses, in the sense that the numbers

$$(12\text{-}5) \qquad prob\ (A_1/B),\ prob\ (A_2/B),\ \ldots,\ prob\ (A_m/B)$$

which represent the agent's degrees of belief in the various hypotheses after the observation may be in much closer agreement with the corresponding numbers for other agents than was the case for the numbers (12-4) which represented his beliefs before the observation. The relevant formula (Bayes's theorem) is easily derived from the probability axioms and the definition of conditional probability: for any particular one of the hypotheses (say, A_1) we have

$$(12\text{-}6) \quad prob\,(A_1/B) = \frac{prob\,A_1}{prob\,B}\,prob\,(B/A_1)$$

$$= \frac{prob\,A_1\,prob\,(B/A_1)}{prob\,A_1\,prob\,(B/A_1) + \ldots + prob\,A_m\,prob\,(B/A_m)}$$

Equation (12-6) holds if neither B nor any of the A's have probability zero.

Example 1: The trick coin

Two agents find a coin lying head up on the floor of a magicians' supply shop and idly start tossing it, without troubling to examine the underside. Initially they agree that either (A_1) the coin has two heads, or (A_2) the coin is normal, so that (12-1) and (12-2) hold with $m = 2$, regardless of whether *prob* is the first or the second agent's belief function; and initially they also agree on the numbers (12-3), where B is any proposition that describes possible outcomes of a series of tosses of the coin. In particular, if B is a complete report on the outcomes of a certain set of n tosses, and if *prob* is either agent's belief function, *prob* (B/A_1) will be 0 if any tails are reported and will be 1 if B says that all tosses yielded heads; and *prob* (B/A_2) will be $1/2^n$ in any case. Thus, if B reports that the outcomes of the first five tosses were *hhtht* we have *prob* (B/A_1) = 0 and *prob* (B/A_2) = 1/32. Now suppose that initially the agents disagree about the probabilities of the propositions A_1 and A_2: the first agent takes these propositions to be equally likely, so that for him, *prob* A_1 and *prob* A_2 are both 1/2, while the second agent takes A_2 to be twice as likely as A_1, so that for him *prob* A_1 is 1/3 while *prob* A_2 is 2/3. If they toss the coin five times and then reconsider the question in the light of the evidence B that is so obtained, they will be in rather close agreement: the numbers *prob*

(A_1/B) and *prob* (A_2/B) will be the same, or nearly so, whether *prob* is the first or the second agent's belief function. In particular, suppose that B reports the first five tosses as yielding *hhtht*. Then by Bayes's theorem (12-6) we have

$$prob\ (A_1/B) = \frac{0}{0 + (prob\ A_2)/32} = 0$$

for either agent, and therefore

$$prob\ (A_2/B) = 1 - prob\ (A_1/B) = 1:$$

the agents are in complete agreement after the observation. On the other hand, if B represents the first five tosses as yielding *hhhhh,* Bayes's theorem gives

$$prob\ (A_1/B) = \frac{prob\ A_1}{prob\ A_1 + (prob\ A_2)/32}$$

so that the numbers *prob* (A_1/B) and *prob* (A_2/B) are

$$\frac{32}{33} \quad \frac{1}{33}$$

for the first agent, and

$$\frac{32}{34} \quad \frac{2}{34}$$

for the second: the agents are very close to agreement after the observation.

12.3 Simple Induction

The class of cases that has received most attention from philosophers goes by the name of *simple induction* or *enumerative induction* or *induction by simple enumeration.* Hume puts the principle in this way (*Treatise,* book 1, part 3, sec. 6):

> *that those instances, of which we have no experience, resemble those, of which we have had experience.*

Hume holds the principle to be sound, argues that its soundness cannot be demonstrated, and assigns it the status of a psychological fact, namely, that we are so constituted as to act as if we were following the principle of simple induction. Thus, the experience of observing many crows, all of which have been black, *causes* me to expect the next crow I observe to be black.

However, the principle is defective in that it fails to specify the sorts of respects in which unexperienced instances are to be expected to resemble experienced instances. Thus, everyone observed so far has been born before the year 2000, but I do not therefore expect that my great-grandchildren will be born before the year 2000. In Nelson Goodman's terminology, some properties are *projectible* while others are not. The property of being black if a crow is projectible; the property of happening before the year 2000 is not. In general, the projectible properties are defined as those to which the principle of simple induction properly applies.

Failure to restrict the principle of induction to the right sort of properties can lead to conclusions which are not only foolish, but actually self-contradictory, as the following version of one of Goodman's examples shows.

Example 2: "It's a goy!"

So far, all boys and no girls have had the *XY* genotype, so that by simple induction there is strong reason to suppose that the first boy born in the year 2000 will have the *XY* genotype, and that the first girl born in the year 2000 will not. Now, define a *goy* as *a girl born before the year 2000 or a boy born thereafter,* and define a *birl* as *a boy born before the year 2000 or a girl born thereafter.* So far, no goys have had the XY genotype, and all birls have, so that by unrestricted simple induction there is strong reason to suppose that the first goy born in the year 2000 will not have the *XY* genotype, and that the first birl will. However, by definition, the first goy born in the year 2000 will be a boy, and the first birl will be a girl, so that the conclusion of a second induction is the proposition that the first boy born in the year 2000 will not have the *XY* genotype, and that the first girl will; and this is the denial of the conclusion of the first induction! This is the case in spite of the fact that although the sentences that express the data for the two inductions are different, they express the same proposition, for the sentence

x is a boy born before the year 2000 and has the *XY* genotype, or is a girl born before the year 2000 and hasn't

is logically equivalent to the sentence

x is a goy born before the year 2000 and does not have the *XY* genotype, or is a birl born before the year 2000 and has the *XY* genotype.

Then if we place no restriction on the resemblances relative to which the rule of simple induction is to be applied, the rule is radically unsound, for it supports contradictory conclusions; and if we amend if by restricting the relevant resemblances to projectible properties, the rule becomes a useless platitude. Then unless the projectible properties can be characterized in some noncircular way, the principle of simple induction is useless, being either unsound or platitudinous.

The most obvious way in which one might hope to get a noncircular characterization of projectibility is to trade on the fact that there are simple terms in English and in other natural languages for such properties as being a boy and being a birl, but that properties like being a goy and being a birl had to be defined in example 2. The point is not that *boy* and *girl* can't be defined in terms of *goy* and *birl:* they can, quite easily. To be a boy is to be a birl born before the year 2000 or a goy born thereafter, and, similarly, a girl is a goy born before the year 2000 or a birl born thereafter. Then one can conceive of a language in which there are simple terms for *goy* and *birl* but not for *boy* and *girl,* in which the latter pair would have to be defined in terms of the former. But the point is that in the languages men actually speak there are simple terms for *boy* and *girl* but not for *goy* and *birl;* that such general facts about actual languages reflect general features of human psychology; and, specifically, that properties for which we have simple terms tend to be properties that we are willing to project. Then existence of a simple term for a property is evidence (but not conclusive evidence) for its projectibility.

Still, the project of characterizing projectibility in these terms seems misguided. The projectibility of certain properties and the nonprojectibility of others are anthropological facts. If the classification were strongly at odds with the way the world is, we should either have perished or altered. Indeed, we keep changing: language evolves, and theories change. Given the belief function of a particular agent, it is a matter of calculation to determine whether a particular predicate is projectible. The belief function summarizes all the relevant data, while the information that the agent's language contains simple terms for such-and-such predicates but not for others gives only some of the data.

I think that all we can legitimately derive from Hume's principle of induction is a definition of projectibility. If we have a sequence of names without beginning or end (decimal expressions for the integers will do) and a predicate (say, *is a boy*) we get a sequence of propositions

$$\ldots, A_{-2}, A_{-1}, A_0, A_1, A_2, \ldots \tag{12-7}$$

attributing the predicate to the named (or numbered) individuals. To say that the predicate is projectible, relative to a belief function *prob,* is to say that for each n, the sequence of numbers

$$prob\ A_n,\ prob\ (A_n/A_{n-1}),\ prob\ (A_n/A_{n-1}A_{n-2}),\ \ldots \tag{12-8}$$

increases toward 1 as a limit. (Consider $n - 1$ as the individual observed just before individual n; $n - 2$ as the one observed just before $n - 1$; and so on.) The situation encountered in example 2 then has the following structure. We are given a predicate (*is a boy with the XY genotype or a girl without it*) which is clearly projectible. Thus, the sequence (12-8) increases toward 1 as a limit, where A_n is the proposition attributing the given predicate to human number n in the sequence (12-7). (For simplicity, imagine that the numbers indicate order of birth, and that there have always been people being born.) We then define another property which gives rise to a sequence of propositions

$$(12\text{-}9) \qquad \ldots, B_{-2}, B_{-1}, B_0, B_1, B_2, \ldots$$

where, in general,

$$(12\text{-}10) \qquad B_n = (A_n)(n < 2000) \vee (\overline{A}_n)(n \geqslant 2000)$$

Now the propositions before B_{2000} in sequence (12-9) are identical with the corresponding propositions in sequence (12-7), while the rest are denials of the corresponding propositions in the other sequence, e.g., by (12-10),

$$B_{2000} = (A_{2000})\,(2000 < 2000) \vee (\overline{A}_{2000})(2000 \geqslant 2000) = \overline{A}_{2000}$$

Thus, the sequence (12-9) can be written

$$(12\text{-}11) \qquad \ldots, A_{-2}, A_{-1}, A_0, \ldots, A_{1999}, \overline{A}_{2000}, \overline{A}_{2001}, \ldots$$

Now if we replace talk about properties by talk about the corresponding sequence of propositions, it is obvious that the sequence (12-11) cannot be projectible if the sequence (12-7) is, e.g., because for $n = 2000$, the fact that the sequence (12-8) increases toward 1 as a limit means that the corresponding sequence of B's will decrease toward 0 as a limit:

$$prob\,(\overline{A}_{2000}/A_{1999}A_{1998} \ldots A_i) \longrightarrow 0 \text{ as } i \longrightarrow -\infty$$

It is indeed a matter of calculation to verify that B corresponds to a nonprojectible property.

It should be borne in mind that where simple induction is not properly applicable, a proposition may still receive probabilistic support of other sorts, as when the proposition $\overline{B}_{2000}$ is strongly supported by the true conjunction of the preceding 2000 A's even though simple induction relative to the B's is inapplicable.

12.4 Confirming Generalizations

Given an endless sequence $A_1, A_2, \ldots$ of propositions, define

$$A(n) = A_1 A_2 \ldots A_n$$

so that $A(n)$ is the conjunction of the first n propositions in the sequence; it is understood that

$$A(1) = A_1$$

We call the sequence "inductive" when *prob* $(A_n/A(n))$ is an increasing function of n which converges to 1 as n increases without bound. There are such sequences, relative to the belief function *prob* of an ordinary agent. It may also be that there are *strongly* inductive sequences relative to *prob:* sequences such that when we define A as the conjunction of all members of the sequence (so that A says that $A_1, A_2, \ldots, A_n, \ldots$ are all true), *prob* $(A/A(n))$ is an increasing function of n, converging to 1 as n increases without bound.

To study the situation, notice that by (5-1) (*f*) we have

$$prob\ A \leq prob\ A(n)$$

for each positive integer n if *prob* satisfies the probability axioms. Then the limit of the sequence

(12-12) $\qquad prob\ A\ (1),\ prob\ A\ (2),\ prob\ A\ (3)\ , \ldots$

is at least as great as *prob A*. However, nothing in the probability axioms implies that the limit of the sequence (12-12) actually equals *prob A*. This must be postulated separately.

(12-13) LIMIT AXIOM. For any positive number e, there is a positive integer n such that each member of the sequence (12-12) after the nth exceeds *prob A* by less than e.

Example 3: Density

For an example of an assignment that satisfies the basic probability axioms but violates the limit axiom, suppose that a positive integer is to be selected at random, and let B_i be the proposition that the selected integer is i. Let the probability field consist of the propositions B_1, B_2, B_3, . . . together with all results of applying the operations of conjunction, disjunction, and denial, where the first two operations may be applied to finite or infinite collections of propositions. Thus the disjunction of all the B's is in the field, and in fact we have

$$T = B_1 \vee B_2 \vee B_3 \vee \ldots.$$

since the selected number must be either 1 or 2 or 3 or . . . Similarly, defining A_i as $\overline{B}_i$ we have

$$F = A_1A_2A_3 \ldots$$

for it is impossible that the selected number be neither 1 nor 2 nor 3 nor

. . . Any proposition other than F in the field can be expressed as the disjunction of a finite or infinite collection of B's. Thus the proposition that the selected number is odd is the disjunction of all B's that have odd subscripts,

$$B_1 \vee B_3 \vee B_5 \vee \ldots$$

and the proposition that the selected number is a power of 10 (1 or 10 or 100 or 1000 or . . .) can be expressed,

$$B_1 \vee B_{10} \vee B_{100} \vee B_{1000} \vee \ldots$$

To determine the probability of a proposition C in the field, express it as a disjunction of B's; for each positive integer n define $C[n]$ as the number of B's with subscripts no greater than n which appear in the disjunction; and define the probability of C as the limit of the sequence

(12-14) $$C[1], C[2]/2, C[3]/3, \ldots, C[n]/n, \ldots$$

if that limit exists. (If not, the probability of C is undefined.) The assignment so defined is called the *uniform density measure on the positive integers*. Thus, if C is the proposition that an odd number is selected, the sequence (12-14) is

$$1, 1/2, 2/3, 2/4, 3/5, \ldots$$

of which the limiting value is 1/2, while if C says that the selected integer is a power of 10 the sequence (12-14) is

$$1, 1/2, 1/3, 1/4, 1/5, 1/6, 1/7, 1/8, 1/9, 2/10, 2/11, \ldots$$

of which the limiting value is 0. Furthermore, for each i we have *prob* $B_i = 0$ since $B_i[n]$ is 0 or 1 accordingly as n is or is not less than i. Then if A_i is the denial of B_i, we have *prob* $A_i = 1$ and, in fact, *prob* $A(n) = 1$ for each finite number n. It follows that the limit of the sequence (12-12) is 1 in spite of the fact that since $A_1A_2A_3 \ldots = F$ we have *prob* $A = 0$. Then here the limit axiom fails in spite of the fact that (as one can show) the basic axioms are satisfied.

Now where the limit axiom is satisfied, we have

(12-15) $$prob\ A = prob\ A_1\ prob\ (A_2/A_1)\ prob\ (A_3/A_1A_2) \ldots$$

where A is the infinite conjunction $A_1A_2A_3 \ldots$ To put matters more neatly, define $A(0)$ as T, and notice that (12-15) identifies *prob* A with a certain infinite product: the product of all numbers of form

$$prob\ (A_n/A(n-1))$$

where n is a positive integer:

(12-16) $prob\ A = prob\ (A_1/A(0))\ prob\ (A_2/A(1))\ prob\ (A_3/A(2)) \ldots$

Since none of the factors in the product (12-16) can be greater than 1, and since typically, all of them will be (positive but) less than 1, one might suspect that the infinite product must be 0. This will be the case if A_i is the proposition that the *i*th toss of a certain coin which the agent takes to be well balanced and fairly tossed will yield a head: we then have $prob\ (A_n/A(n - 1)) = 1/2$ for each n, and the probability that all tosses yield heads will be

$$prob\ A = (1/2)(1/2)(1/2) \ldots = 0$$

But in general, the infinite product (12-16) need not be 0. Thus the product

(12-17) $$(1/2)(3/4)(7/8)(15/16) \ldots$$

in which the *n*th factor is $(2^n - 1)/2^n$ has a positive value: successive factors approach 1 fast enough to overbalance the tendency of an infinite product of positive numbers less than 1 to go to 0. It can be shown that the product (12-16) is positive if and only if none of its factors are 0 and in addition the infinite sum

(12-18) $prob\ (\overline{A}_1/A(0)) + prob\ (\overline{A}_2/A(1)) + prob\ (\overline{A}_3/A(2)) + \ldots$

has a finite positive value. Thus, the product (12-17) satisfies the conditions; in particular, the infinite sum (12-18) is then

$$\frac{1}{2} + \frac{1}{4} + \frac{1}{8} + \frac{1}{16} + \ldots = 1$$

If *prob A* is positive, and only then, the generalization *A* is confirmable by simple induction (relative to *prob*) in the sense that $prob\ (A/A(n))$ can be brought arbitrarily close to 1 by making n large enough; for *prob A* is the limit that $prob\ A(n)$ approaches as n increases without bound, and we have

$$prob\ (A/A(n)) = \frac{prob\ A}{prob\ A(n)}.$$

Similarly, *A* is confirmable by Bayes's theorem if and only if *prob A* is positive. Then if we know an agent's belief function at a certain time, we know (in principle) what generalizations he takes to be confirmable at that time: they will be precisely the generalizations to which his belief function then assigns positive probability. It is only to such generalizations that his observations can lead him to attribute positive probabilities, as long as he confines himself to the method of conditionalization or the methods of chapter 11 for modifying his belief function in response to experience.

And as long as he confines himself to such methods, no amount of experience can lead him to attribute the slightest degree of belief to a generalization in which his initial degree of belief is zero.

But in the absence of special reasons to the contrary, it is to be supposed that the agent's degree of belief in a universal generalization will be zero; for willingness to attribute positive probability to a universal generalization is tantamount to willingness to learn from experience at so great a rate as to tempt one to speak of "jumping to conclusions."

Example 4

If *prob* $(A_n/A(n-1))$ is $(2^n-1)/2^n$ for each positive integer n, the probability of the infinite conjunction $A = A_1A_2A_3 \ldots$ is given by (12-17) and is positive: the agent regards A as confirmable. But he is also prepared to bet very heavily on the next individual's having the property in question after having seen rather few individuals, provided they have all had the property. Thus, having observed nine individuals and seen that $A(9)$ is true, his degree of belief in A_{10} will be greater than .999: he is willing to risk $999 in order to gain one dollar if A_{10} is true.

In general, a belief function *prob* might be said to permit learning from experience or induction relative to the sequence $A_1, A_2, A_3, \ldots$ when the following equivalent conditions hold for all positive n:

$$\text{(12-19)}\quad \text{(a)}\ \frac{prob\,(A_{n+1}/A(n))}{prob\,(A_n/A(n-1))} > 1 \qquad \text{(b)}\ \frac{prob\,(\overline{A}_{n+1}/A(n))}{prob\,(\overline{A}_n/A(n-1))} < 1$$

Thus, the more more members of the sequence of A's are observed to be true without exception, the more likely is it thought that the next member will be true. Similarly, we might say that *prob* permits *strong* induction relative to the A's if the more members of the sequence are observed to be true without exception, the more likely is it thought that all members of the sequence are true. Aptness of this terminology is shown by the fact that *prob A* is positive if and only if the sum (12-18) is positive and none of its terms are 1. Thus, applying Cauchy's ratio test to (12-18), we find the following to be a sufficient condition for strong induction to be possible relative to $A_1, A_2, A_3, \ldots$: there is a constant a such that for each positive integer n we have

$$\text{(12-20)}\qquad \frac{prob\,(\overline{A}_{n+1}/A(n))}{prob\,(\overline{A}_n/A(n-1))} \leqslant a < 1$$

Comparing (12-20) with (12-19) (b) we have an illustration of the thesis that the strong sort of induction is simply a high degree of the other.

The connection between induction and strong induction has an important bearing on the interpretation of probability statements in terms of willingness to bet. On this interpretation, the statement that an agent has degree of belief p in the generalization A means that he is willing to bet on A in such a way as to pay p dollars in order to get one dollar if A is true, and get nothing if A is false. But how could one know that such a bet had been won? No matter how large a finite segment of the sequence $A_1, A_2, A_3, \ldots$ has been observed to consist exclusively of truths, it is always possible that the next member will prove false and that therefore the bet will be lost. Then no finite sequence of observations can make it quite clear that the agent should be paid the dollar that is his due if A is true. The agent would be foolish, to bet on a universal generalization; and therefore the interpretation of the statement "$prob\ A = p$" in terms of willingness to bet seems to fail.

But the connection between *prob* A and the sequence (12-12) saves the betting interpretation. The fact that the probability of the infinite conjunction A is p will be reflected in the fact that the probabilities of the finite conjunctions $A(1), A(2), A(3), \ldots$ converge to p. It is quite possible and reasonable to bet on any one of these finite conjunctions; and it is equally possible and reasonable to bet on all such finite conjunctions simultaneously. (Bet 1/2 dollar on $A(1)$, 1/4 dollar on $A(2)$, and in general, $1/2^n$ dollars on $A(n)$, so that no matter how many bets are won, a bank of one dollar suffices to guarantee all payments. By making a sufficiently large finite number of observations, the unrealized potential gains to the agent from the infinity of unresolved bets can be made as small as he pleases.) The riskiness of assigning positive probability to a universal generalization A is also reflected in this interpretation, for if *prob* A is positive, the ratio $prob\ A(n + 1)/prob\ A(n)$ can be brought arbitrarily close to 1 by choosing n high enough, and therefore there will be a single conditional bet on A_{n+1}, conditionally on $A(n)$, which seems advantageous to the agent but which is as risky as you please in the sense that the amount p the agent is willing to pay in order to get a dollar if $A(n + 1)$ is true—and get nothing if $\overline{A}_{n+1}A(n)$ is true, and get his p dollars back if $A(n)$ is false—is as close to 1 as you please; for the bet will appear advantageous to the agent if p is even slightly less than $prob\ (A_{n+1}/A(n))$, which is $prob\ A(n + 1)/prob\ A(n)$.

12.5 Objectivity and Learning

In the case of universal generalizations, we have seen that an adequate account of the nature of an agent's belief in one proposition $(A_1A_2A_3 \ldots)$ may involve an account of the subjective probabilities that he attributes to an infinite sequence of other propositions $A(1)$, $A(2)$, $A(3)$, . . . The same remark can be used to solve an old puzzle which has been posed again by K. R. Popper in a version which he calls "the paradox of ideal evidence."

The puzzle is this. Sir Karl confronts you with an oddly shaped coin. It may be fair or it may be biased; you do not know. That is, you do not know whether the objective probability of tossing a head is 1/2. Sir Karl gives you some money and asks for the least favorable odds at which you are willing to bet it on the proposition A_n that the nth toss of the coin will yield a head. I suppose the odds you propose are 1:1, which means that your belief function *prob* assigns the value 1/2 to A_n.

Now you are invited to experiment with the coin, and after a while you conclude that the coin is perfectly fair: you conclude that the objective probability of tossing a head with it is 1/2. Again Sir Karl asks for the least favorable odds at which you are willing to bet on A_n, and again you reply 1:1. In other words your new subjective probability measure *PROB* again assigns the value 1/2 to A_n.

But this is very strange (says Sir Karl): you have gone from a state of complete ignorance of the objective probabilities to a state of perfect knowledge of them, and yet the value your subjective probability measure assigns to A_n has not changed.

Here you might ask: What should I have done? When you so kindly let me experiment with the coin I found that the odds I originally offered were (luckily) precisely the ones I should have offered had I known the objective probabilities. I have more reason now than before for proposing odds of 1:1.

Sir Karl: Precisely my point! Your new subjective probability measure *PROB*, should incorporate your new knowledge. It should reflect the fact that you now know that 1:1 are eminently fair odds on A_n. But *PROB* assigns the same value to A_n that *prob* did.

Now let me intervene. I suggest that at this point your attention is being misdirected to A_n as a proposition to which your old and new belief functions must surely assign different values, if there is any such proposition. But as you both agree, *prob* and *PROB* should both assign the value 1/2 there, so that can not be the locus of the difference. Nevertheless, there is a difference: *prob* and *PROB* will assign different values to any proposition $A(n)$ that asserts, concerning $n \geqslant 2$ distinct tosses, that all of them yield heads. To any such proposition, *PROB* assigns the value $1/2^n$; but to the same such proposition *prob* must assign a higher value, if you hope to learn from experience.

You can bet not only on the individual propositions A_1, A_2, A_3, . . . which assert that the first toss yields a head, that the second toss yields a head, that the third toss yields a head, and so on, but also on any truth functional compound of those propositions; for example, on the proposition $A_1A_2 \vee A_2A_3 \vee A_1A_3$, which asserts that at least two of the first three tosses yield heads. To determine the value *prob* assigns to all such propositions it is not sufficient to know merely the values it assigns to the individual *A*'s: in addition, one must know the values *prob* assigns to all conjunctions of the *A*'s taken two at a time, three at a time, and so on. For each finite set of trials, one must know the value that *prob* assigns to the proposition that all of those trials yield heads.

Some simplification is possible, however, since we may take *prob* to be a *symmetric* probability measure. This means that *prob* treats all tosses on a par, assigning the same value, 1/2, to each of the *A*'s and, furthermore, assigning to any conjunction of *n* of the *A*'s a value that depends only on *n*, and not on which particular *A*'s were involved. Thus, for $n = 3$, *prob* assigns the same probability to the proposition $A(3)$ that the first three tosses all yield heads that it does to the proposition $A_5A_{50}A_{500}$ that tosses number 5, 50, and 500 all yield heads. Then *prob* is completely determined for all truth functions of the *A*'s by the infinite sequence of numbers (12-12): by the values it assigns to the propositons $A(1)$, $A(2)$, $A(3)$ . . . When you decided to offer odds of 1:1 on A_n, you fixed the value of *prob* $A(1)$ at 1/2; but since the values of *prob* $A(n)$ for $n = 2, 3, 4, \ldots$ were undetermined, *prob* was not completely defined.

When you choose the numbers *prob* $A(n)$ for $n = 1, 2, 3, \ldots$, you choose the rate at which you can learn from experience, for by definition, *prob* $(A_{n+1}/A(n))$ is the ratio of *prob* $A(n + 1)$ to *prob* $A(n)$. Since we know the value *prob* $A(1) = 1/2$, the rest of the numbers *prob* $A(n)$ will be determined by the values assigned to the ratios of successive numbers in that sequence; and this means that the sequence *prob* $A(n)$ with $n = 1, 2, 3, \ldots$ is determined by its first member and by the sequence

(12-21) $\quad prob\ (A_2/A(1)),\ prob\ (A_3/A(2)),\ prob\ (A_4/A(3)),\ \ldots$

which defines the rate at which you propose to learn from experience, given various amounts of evidence.

Now it may not be difficult to determine your belief function *prob*. Imagine that you are watching successive tosses of an odd-looking coin; imagine that each toss, in turn, yields heads; and ask yourself what odds you would think fair, at each stage of this procedure, on the next toss yielding a head. (In the situation described here, you might eventually come to suspect, more and more strongly, that the coin has two heads.)

*Example 5: The belief function m**

It might be reasonable to offer odds of $n:1$ on the *n*th toss yielding heads, conditionally upon all previous tosses having yielded heads. This means that the ratio between *prob* $A(n + 1)$ and *prob* $A(n)$ will be

(12-22) $$prob\ (A_{n\omega 1}/A(n)) = \frac{n + 1}{n + 2}$$

so that since *prob* $A(1)$ is 1/2 the sequence of numbers of form *prob* $A(n)$ will be

(12-23) $$1/2,\ 1/3,\ 1/4,\ \ldots,\ 1/(n + 1),\ \ldots$$

The function *prob* so determined is identical with the function m^*, which at one time Carnap considered using as a basis for inductive logic, and which W. E. Johnson had earlier proposed (in *Logic,* vol. 3 [1924; reprint ed. New York: Dover, 1964], appendix).

In *The Continuum of Inductive Methods,* Carnap explores a certain class of symmetric probability measures and proposes that one choose a single measure from this class in terms of the rate at which one wishes to learn from experience, bearing in mind that if the rate is high, there is a high risk of jumping to false conclusions. (Concerning practical choice of λ, see section 5.3 of I. J. Good, *The Estimation of Probabilities* [Cambridge: MIT Press, 1965]. Carnap's parameter λ is Good's flattening constant k. Here again, W. E. Johnson had proposed the same methods much earlier: *Mind* 41 [1932]: 408–23.) The chosen measure is to play the role of an idealized initial belief function which would be appropriate for an intelligent agent who somehow had no experience whatever; the agent's actual belief function then ought to be derivable from the initial belief function by conditionalization or by the methods of chapter 11. It is just this process of choosing a reasonable belief function in terms of the rate at which one wishes to learn from experience that is illustrated in example 5. To translate that example into Carnap's terms one must abstract: forget about the coin, and about the existence of an objectively correct probability measure. Interpret A_1, A_2, A_3, . . . as propositions that attribute some property successively to the first individual, the second individual, the third individual, etc., in the universe of discourse. Imagine that the individuals have been ordered by some secret process so that you know nothing useful about an individual when you know its position in the sequence. It is then reasonable to treat all the A's on a par and to adopt a symmetric probability measure as your initial belief function.

The general situation that Carnap treats is more complicated than this. Our coin-tossing example is the analogue of a language containing only one primitive predicate; but in *The Continuum of Inductive Methods,* Carnap studies languages having arbitrary finite numbers of monadic primitive predicates; and in subsequent work he extends the scope of his inductive logic still further.

12.6 De Finetti's Representation Theorem

De Finetti's advice about choosing a belief function in connection with coin-tossing uses a different point of view which supplements Carnap's: instead of asking for the rate at which you hope to learn from experience, he asks (in effect) for the odds at which, initially, you would bet that the objective probability of heads lies in various intervals from 0 to 1. I say "in effect" because de Finetti does not believe that there are such things as objective probabilities. However, he does believe that there are such things as subjective probability measures, and he believes that in the coin-tossing example, given enough time to experiment with the coin and to think about it, different people's subjective probability measures would all pretty much agree. Then if we substitute some such notion as the subjective probability measure that in fact we should all come to in the end for the notion of an objective probability measure, we can fairly describe de Finetti's approach in the terms we have just used.

In particular, de Finetti asks you concerning each number from 0 to 1 for the odds at which you would bet that the objective probability of heads is no greater than that number. That is, if r is the objective probability of heads, he asks for the function F, such that $F(x)$ is the subjective probability that r is no greater than x.

One reasonable choice of F might correspond to the situation that is loosely described by saying that all possible values of r are equally likely. The graph of F would then be as shown in figure 12.1. On the other hand, to indicate that we regard all values other than 1/2 as equally likely, whereas we would bet at even money on 1/2, we could draw the graph of figure 12.2 for F. Now de Finetti proposes that you choose as your subjective probability measure, *prob,* a certain *mixture* or *weighted average* of the various possible objective probability measures. The proportions in which different objective probability measures are represented in this mixture—the weights attached to them in the average that de Finetti wants you to form—are determined by the function F.

If we symbolize the objective probability measure in which the probability of heads is r by obj_r, then the general mathematical expression for the mixture will be given by the following formula, in which the subjective probability of an arbitrary proposition B concerning outcomes of tosses of the coin is expressed as a weighted average of its various possible objective probabilities:

$$prob\ B = \int_0^1 obj_r B\ dF(r). \qquad \text{(12-24)}$$

(Obj_r will be the symmetric probability measure determined by the condition that $obj_r\ A(n) = r^n$. Then $prob\ A(n) = \int_0^1 r^n\ dF(r)$, so that *prob A*

Fig. 12.1

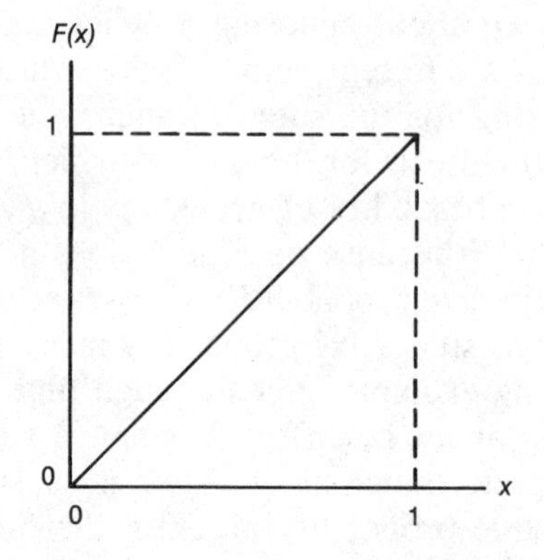

(n) will be the nth moment of the cumulative distribution F.) In certain special cases the expression (12-24) can be greatly simplified. Thus, if initially you are convinced that the objective probability must have one of the values in a certain sequence $r_1, r_2, \ldots$, and if you regard the probabilities of these successive values as being $p_1, p_2, \ldots$, then your subjective probability measure will be given by a weighted sum,

$$\text{(12-25)} \qquad prob\ B = p_1 obj_{r_1} B = p_2 obj_{r_2} B + \ldots$$

The graph of the function F in this case would be a staircase, where the p's are the heights of the successive steps.

When the graph of F is the straight line shown in figure 12.1, formula (12-24) determines the same function *prob* that was suggested in example 5, namely, Carnap's measure function m^*. From the point of view of example 5, m^* is the function you get when you decide to offer odds of n:1 on heads on the nth toss, conditionally upon all earlier tosses having yielded heads. Using de Finetti's technique we find the same function resulting from the initial assumption that all possible values of the objective probability of heads are equally likely. Thus, combining the two approaches, we find that any decision about the rate at which you hope to learn from experience corresponds to some initial set of estimates about where the objective probability is likely to lie, in the interval from 0 to 1. And conversely, any such set of estimates determines the rate at which you can hope to learn from experience. (Note that Carnap had something like this double view of the choice of a measure function: the point of view taken in *The Logical Foundations of Probablity* is like that I attribute to de Finetti, while the point of view in which one chooses the rate at which one hopes to learn from experience is that of *The Continuum of Inductive Methods*.)

To illustrate some of the ramifications of de Finetti's approach, let us conclude by applying it to Popper's question, "How is your knowledge

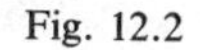

Fig. 12.2

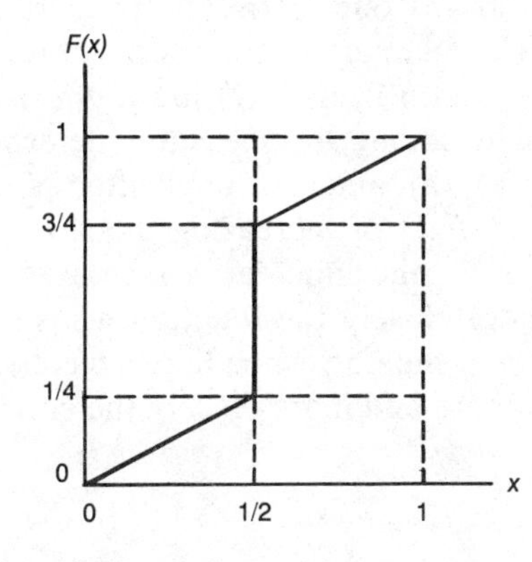

of the objective probabilities reflected in your subjective probability measure?'' Compare the three possible shapes of the graph of F that are shown in figure 12.3. The first shape corresponds to a state of belief in which all possible values of the objective probability have equal subjective probability. The corresponding *prob* is identical with Carnap's m^*. The third corresponds to the opinion that the objective probability is certainly r; and the second indicates an intermediate degree of certainty about the objective situation. The three graphs represent three possible degrees of sharpness of the agent's estimate of the objective probability: in figure 12.3 (a), sharpness is minimum; in figure 12.3 (b), intermediate; and in figure 12.3 (c), maximum. Corresponding to these three graphs are three symmetric subjective probability measures, the first being appropriate for a person who is totally ignorant of the objective probabilities, the second being appropriate for someone who has a little knowledge, and the third being appropriate for someone who is completely informed.

Fig. 12.3

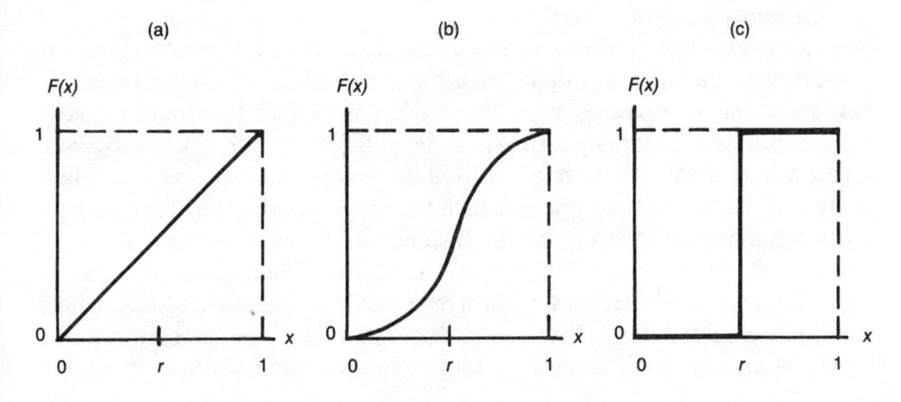

De Finetti has shown that as one acquires more and more statistical evidence about the coin—as one learns the proportion of heads in the first n tosses for larger and larger n—one's a posteriori belief function as derived by conditionalization from (say) the a priori belief function m^* moves from left to right among the measures determined by the three graphs of figures 12.3 (a), (b), and (c). Acquisition of statistical evidence about the coin has the effect of increasing the slope of the graph of F near some value $x = r$. In the limit, as n increases indefinitely, F approaches more and more closely the function whose graph is a vertical straight line at $x = r$, and the corresponding subjective probability measure *prob* approaches more and more closely the correct objective probability measure obj_r.

12.7 Objectification

To believe in the existence and utility of a subjectivistic notion of probability is not to deny the existence or utility of a parallel objectivistic probability concept as well. Indeed, as we have seen, important features of subjective probability assignments are sometimes most naturally interpreted (via de Finnetti's representation theorem) as reflecting the agent's beliefs about the objective probabilities of certain propositions. On the other hand, it is attractive to have a single basic notion of probability: in the interest of conceptual economy one might wish somehow to reduce the objectivistic concept to the subjectivistic one.

I shall now propose a reduction of a sort. It will be suggested that the objectivistic notion is obtained by adding an extra feature to the subjectivistic notion, and that the resulting *objectification* of the concept of probability has a parallel for the concept of desirability. Let us begin by considering one of the things possibly meant by speaking of objective desirabilities, relative to an agent.

Example 6: Whale steak

In an experimental mood, the agent decides to try whale steak for dinner: it is cheap and might taste like beefsteak, which he likes. Of course, it might prove to have a fishy flavor, which he would heartily dislike; but he takes this possibility to be sufficiently improbable to make trying whale steak the best of the available acts. Simplifying, we might represent the situation as one in which the proposition A that he has whale steak for dinner is taken to be the disjunction of two incompatible propositions, B (that he has whale steak for dinner and that whale steak tastes like beefsteak) and C (that he has whale steak for dinner and that whale steak has a fishy flavor). Either B or C contains the objective truth about whale steak, insofar as the truth bears on his enjoyment of dinner, so that one might wish to say that the objective desirability of whale steak for dinner is either *des B* (the subjective desirability of whale steak that tastes like beefsteak) or *des C* (the subjective desirability of whale steak that tastes fishy); that experience will show the agent which of these is the actual objective desirability of whale steak for dinner; and that *des A,* the subjective desirability of whale steak for dinner, is a weighted average of the values that the agent thinks the objective desirability might have, in which the weights are the probabilities that the objective desirability has the various values.

The general scheme that is illustrated by example 6 can be described as follows. Before determining whether or how the proposition A is true, the agent partitions A into a sequence of pairwise incompatible propositions A_1, A_2, A_3, etc., whose disjunction is A. Each member A_n of the partitioning represents a way in which A might prove true. In the interesting cases, *des* A_n will have one value or another, depending on n; and each member A_n of the partitioning will be highly homogeneous relative to the agent's belief function *prob,* in the sense that if B_n is a proposition in the agent's preference ranking which implies A_n (and is thus a way in which A_n might be true), the desirability of B_n will be very nearly equal to the desirability of A_n unless the probability of B_n is very much less than that of A_n. Then for A_n to be homogeneous means that barring contingencies which the agent takes to be highly improbable, one way of A_n's proving true is pretty much as good as another, in the agent's estimation. (If the agent knew that whale steak tasted fishy, he would inquire no further; he would be no more pleased to have it taste like haddock than like flounder.) *Relative to the given partitioning, the agent identifies the objective desirability of A with the subjective desirability of that unknown member of the partitioning which is in fact true.*

The desirability assignment *des* is subjective in the sense that for any proposition A, the facts which are relevant to the determination of *des A* are facts about the agent; they are not facts that bear on the truth of A. On the other hand, it is characteristic of magnitudes which are said to be objective features of the objects to which they are attributed that the facts which are relevant to their determination are facts about those objects. (Apparently, a well-defined subjective property of an "external" object is an objective property of the perceiving subject or of the ordered pair consisting of the subject and the object, so that the objective/subjective distinction corresponds to the distinction between direct and oblique forms of speech.) The function of the partitioning of A into A_1, A_2, A_3, etc., in objectifying the notion of (subjective) desirability is to specify the external facts which are relevant to the determination of the (now objective) magnitude. Of course, the magnitude is objective relative to the given subjective desirability assignment: relative to the agent in question. Then strictly speaking, objective desirability is a function of two entities: an agent and a proposition. There are two sorts of facts about the agent which are relevant to the determination of the objective desirability of a proposition, relative to him. One sort of fact is given by his subjective desirability assignment, the other by his partitioning of the proposition in question. Thus, we need to know three things in all in order to determine the objective desirability of a proposition A: the agent's partitioning of A, the subjective desirabilities of the members of the partitioning, and the facts as to which member of the partitioning is true.

It is instructive to consider a further complication of this scheme, which suggests itself in cases where there is some continuously variable magnitude (say, the temperature of the water) which is relevant to the agent's anticipated enjoyment of the truth of some proposition A (say, that he plunges into the pool). Here there is no finite or infinite sequence A_1, A_2, A_3, etc., which can exhaust the relevant differences between the ways in which A might come true (although in practice, some large finite sequence would surely do). To do full justice to the theoretical possibilities we must regard any such sequence as only an approximation to the full partitioning of A in which we are interested—a partitioning which has as many members as there are real numbers. However, we must regard the ultimate partitioning as being determined by an infinite sequence of finite partitionings.

Example 7: Taking the plunge

The agent considers that the temperature of the pool might have any value from 0 to 100 degrees Centigrade. For a first approximation, he partitions A (the proposition that he takes the plunge) into $A^1_1, A^1_2, \ldots, A^1_{10}$ where A^1_n is the proposition that he takes the plunge and the temperature of the water is between $10(n - 1)$ and $10n$ degrees Centigrade. For a second approximation he splits each member of the first partitioning into ten propositions, thus obtaining $A^2_1, A^2_2, \ldots, A^2_{100}$, where A^2_n is the proposition that he takes the plunge and the temperature of the water is between $(n - 1)$ and n degrees Centigrade. In general, for each positive integer m, he envisages a partitioning of A into 10^m propositions A^m_n ($n = 1, 2, \ldots, 10^m$), where A^m_n is the proposition that he takes the plunge and the temperature of the water is greater than $100(n - 1)/10^m$ but not greater than $100n/10^m$ degrees Centigrade. He then defines the objective desirability of A as the limit of the sequence of subjective desirabilities of the true members of the successive partitionings, if such a limit exists.

The partitionings that were used to define the objective desirability of A in example 7 have two essential characteristics. First, the probability *prob* A_n^m of any member of any partitioning is positive. Second, the partitionings in the sequence become indefinitely *fine* in the following sense. We define the *coarseness* of the *m*th partitioning,

$$A_n^m, A_2^m, A_3^m, \ldots$$

to be the smallest number which is at least as great as any of the numbers

$$prob\ A_1^m, prob\ A_2^m, prob\ A_3^m, \ldots$$

Now to say that the partitionings become indefinitely fine means that for any positive number *e*, no matter how small, there is a positive integer *m* such that the coarseness of all partitionings after the *m*th is less than *e*.

We objectify the notion of probability in a rather similar way. To determine the objective probability of a proposition *A*, relative to an agent, we need to know his subjective probability measure, *prob;* we need to know a sequence of partitionings of the necessary proposition *T*, which have the two characteristics stated in the preceding paragraph; and for each partitioning in the sequence, we need to know which member of it is true. (To determine the objective probability "approximately," it is sufficient to know which members are true in some finite number of the partitionings; but there is no guarantee that such an "approximation" will be close to the actual objective probability.) We then define the objective probability of *A* as the limit (if any) of the values of

$$prob\ (A/B[m])$$

as *m* increases without bound, where *B*[*m*] is the true member of the *m*th partitioning. Note that the square brackets have a different significance here from what they have in (12–14) above.

Since the choice of the infinite sequence of partitionings may seriously affect the value of the objective probability of *A* (as may the choice of the subjective probability measure *prob*), it is clear that objective probability as we have defined it here may be a highly subjective magnitude in the sense that different agents may apply the definition and emerge with different objective probabilities for a proposition *A*, where the differences are not of the sort that can be resolved by determining the truth values of a larger set of propositions in the agents' common probability field. However, I take this situation to correspond to the fact, stressed by all objectivists, that not all propositions have objective probabilities in the usual sense of the term. Typically, objectivists will hold that it is only such propositions as can be placed in a natural way as members of a sequence of propositions that have objective probabilities: propositions of the form, the *n*th trial of this experiment will be successful. Where *A* is the proposition that a certain horse will win a certain race, there is some difference of opinion as to whether one can speak of an objective probability. In such cases, objectivists are sensitive to the possibility of placing A in many different series, with equal claim to naturalness, where the limiting relative frequency of truths in the different series differs (fillies, two-year olds, horses that have run at this track). One might consider forming the "product" of all such series (two-year-old fillies that have run at this track), except that such product series tend to be not only finite but distressingly short.

But in cases where frequentists agree, there is some preeminently natural series within which A can be placed, and a generally accepted sequence of partitionings (relative frequencies in the first n trials, for $n = 1, 2, \ldots$) representing what all agents can be expected to have in mind; and there is sufficient agreement between the belief functions of different agents so that whatever differences may initially exist between them will be swamped out by the influx of data. (See remarks concerning figure 12.3.) Thus, if *prob* and *PROB* are the belief functions of two agents, the cases in question are of the sort where for all rather large but finite m, *prob* $(A/B[m])$ will differ at most very slightly from *PROB* $(A/B[m])$.

Example 8: Coin tossing

Suppose that the belief function *prob* is symmetric, and that the graph of the distribution function F has no jumps: thus, that it has some such shape as is shown in figures 12.3 (a) or (b), not in figure 12.3 (c) or figure 12.2. The basic sequence of partitionings that agents would naturally use in defining the objective probability of the proposition A_i that the ith toss of a certain coin yields a head would be obtained by reference to the relative frequencies of heads among the first m tosses, for $m = 1, 2, 3, \ldots$ Thus, the mth partitioning would consist of the $m + 1$ propositions

$$B_0^m, B_1^m, B_2^m, \ldots, B_m^m$$

where B_n^m asserts that there were n heads in the first m tosses. Alternatively, one might use a sequence of partitionings which refer to the particular outcomes of particular tosses: the mth partitioning would be

$$C_1^m, C_2^m, \ldots, C_M^m$$

where $M = 2^m$ and where the different C^m describe the 2^m different possible sequences of heads and tails on the first m tosses in the sequence consisting of tosses number

$$1, 2, \ldots, i - 1, i + 1, \ldots$$

(Toss number i is excluded since otherwise we should have *prob* $(A_i/C[m])$ either 0 or 1 for all $m \geqslant i$, depending on whether A_i is false or true.)

The first sequence of partitionings in example 8 illustrates the fact that the partitionings need not be nested. The second sequence raises an interesting question: Shouldn't we include some requirement which codifies the principle on which the *i*th toss is excluded from the reckoning? I think not, for it is one of those points upon which objectivists may profitably be endorsed in differing: whether the objective probability of getting a head on the first toss of a well-balanced coin that has already been tossed should be taken to be 1/2 or 0 (if the first toss turned up tail) or 1 (if the first toss turned up head). One can have it either way; and since differences of judgment between alternative sequences of partitionings that yield different objective probabilities is a pervasive feature of objectivistic practice, it is well to allow for it in the definition.

One final point: in the determination of objective probabilities as in the determination of inductive probabilities it may well make a difference what hypotheses have been proposed or thought of, and it is impossible to think of all hypotheses in advance. In the objective case the hypotheses will most directly affect the question of which sequences of partitionings are thought acceptable. In the inductive case the hypotheses most directly affect the choice of a belief function, as when one wishes to consider the possibility that a certain universal generalization may be true.

Example 9: McBannister's hypothesis

Putnam and McBannister are idly flipping a coin. The first seventeen tosses have the following results:

hhhthththttththttth

Both men agree that the coin is likely to be fair, since about half of the tosses have been heads; but suddenly McBannister notices a pattern: it is the prime numbered tosses, and only they, that have turned up heads! Putnam dismisses it as a curious coincidence, but McBannister believes the coin is being influenced by the spirit of his deceased grandfather, a keen number-theorist, and accordingly hypothesizes that the pattern will continue indefinitely. If Putnam wishes to test McBannister's hypothesis by the methods we have been considering, he must alter his (previously symmetric) belief function so as to attribute a certain positive (if minute) subjective probability to the proposition that all and only the prime-numbered tosses will yield heads. He must then use some such partitioning as the second in example 8 in evaluating the objective probability of A_i: the first partitioning will not do because it throws away data that are essentially relevant to McBannister's hypothesis. (If that hypothesis is correct, the objective probability of A_i will be 1 or 0 accordingly as i is or is not a prime, and this will be so even though the *i*th toss has been excluded from the sequence upon which the partitionings were based.)

12.8 Conclusion

In this chapter I have tried to indicate some of the directions in which the concepts of subjective probability and desirability might be extended beyond their native ground in the theory of individual decision-making, into the realms of the objective and the intersubjective. The treatment has been incomplete and partly conjectural, and the matters treated, while important, have been peripheral to the main topic of the book. Thus, it seems best to close on a different note, reiterating the central theme of the book: that in a certain sense, the Bayesian account of deliberation provides a logic of decision.

We may imagine that the propositions with which the agent is immediately concerned are expressible in some idealized language: idealized in the sense that the (declarative) sentences of that language have fixed (although often unknown) truth values, independent of the contexts of their utterance. (Sentences like "It's hot" (here, now) are replaced by sentences like "It's hot in Princeton, N.J., at noon on April 20, 1965.") We might perfectly well take such sentences to be the objects that are ranked by the agent's preference ordering; but for consonance with the more common way of speaking in mathematical probability theory, we use the corresponding propositions instead. Here propositions (the *events* of mathematical probability theory) are identified with certain sets of sets of sentences as follows. Let us call a nonempty set of sentences a *novel.* The novel is *consistent* if the sentences that constitute it do not logically contradict each other; and a novel is *complete* if no sentence of the language could be added to it without rendering it inconsistent. Equivalently, a complete, consistent novel is a set of sentences which contains exactly one of each contradictory pair of sentences in the agent's language.

A complete, consistent novel describes a *possible world* (an *elementary event,* in the terminology of mathematical probability theory) in as much detail as is possible without exceeding the resources of the agent's language. But if talk of possible worlds seems dangerously metaphysical, we can focus attention on the novels themselves and speak of a complete, consistent novel as actually being a possible world. (This sort of abuse of language is characteristic of mathematical talk and is harmless, since anything we wish to say about a possible world can be said quite as well by making a corresponding statement about a complete consistent novel.) Then possible worlds are complete consistent novels, which in turn are certain sets of sentences. It makes sense to speak of the *set* of sentences as a novel because we have assumed that the sentences of the agent's language have fixed truth values (fixed meanings, I would as soon say) regardless of context. Then the order in which we read the sentences of such a novel makes no ultimate difference: sooner or later, any question about the possible world will be answered. The question, "Is sentence S true in the world that the novel describes?" will eventually be answered yes if S itself is in the novel, and will otherwise be answered no, when we finally read the denial of S in the novel.

Now propositions are identified with sets of possible worlds. In particular, the proposition expressed by a sentence S of the agent's language will be identified with the set of possible worlds that contain S. In general, a proposition is identified with the set of possible worlds in which (as we would ordinarily say) that proposition would be true. It is to be expected that for certain sets of possible worlds there will be no corresponding sentence in the agent's language: no sentence that belongs to all and only the possible worlds in the set. The propositions are more numerous than the sentences if we admit all sets of possible worlds as propositions. But for present purposes we may imagine that the preference ranking contains only propositions that are expressed by sentences of the agent's language.

An unknown one of the possible worlds is actual: an unknown one of the complete consistent novels is true. Indeed we know some things about the true novel: it belongs to every true proposition, and we know what some of those are. (The actual world is identified with the set of all true sentences of the agent's language, and we know of certain sentences that they are true.) But everything we know is compatible with any of an infinite number of possible worlds' being actual.

If $P_1, \ldots, P_n$ and C are propositions, then $P_1, \ldots, P_n$ as premises deductively imply the conclusion C if and only if every possible world that belongs to all n of the premises belongs to the conclusion as well. Then in particular, if the actual world belongs to all of the premises, it must belong to the conclusion as well: if all premises are true, the conclusion must be true. But the inference is valid in such a case even if the actual world fails to belong to all of the premises. Deductive logic uses the notion of *true-in-world-w* where "*w*" is a variable, but it does not use the unqualified notion of *truth:* of *true-in-world-a* where "*a*" denotes the actual world. Nothing in logic depends on which novel describes the actual world.

We can define a certain probability assignment, $prob_a$, by saying that for any proposition A, $prob_a\,A$ is 1 or 0 accordingly as the actual world does or does not belong to A. And more generally, if w is any possible world, $prob_w$ assigns 1 to the propositions that contain w, and assigns 0 to the propositions that do not. For any possible world w, $prob_w$ is indeed a probability measure in the sense that it satisfies the probability axioms (5-1); and it would be highly desirable, if it were possible, for the agent to use $prob_a$ as his belief function. It would then be of no moment that his preference ranking would violate the hypotheses of the uniqueness theorem, for he would have no occasion to make decisions in the face of uncertainty; nor would he have any use for the processes of objectification described in this chapter, or for the methods of probability kinematics described in the preceding one.

But as matters stand, an agent's belief function will be some sort of average of the probability measures $prob_w$, for all possible worlds w: *prob* A will be his degree of belief in the thesis that the unknown actual world belongs to A. Any possible proposition which is expressible in the agent's language might, for all he knows, come true in various ways, which he might value differently. Thus, the proposition that the agent inherits $10,000 next year will be more or less gratifying in the event, depending on whose death proves to have yielded the gain, and this variability may have a strong influence on the desirability he attributes to that proposition. If he thinks it highly unlikely that he is anyone's heir but his brother's, then he takes the world to be such that the prima facie good proposition, that he inherits $10,000 next year, is related to a prima facie bad proposition, that his brother dies this year or next, in such a way that, all things considered, the desirability of the former proposition is nearly the same as the desirability of its conjunction with the latter. In the present theory, desirabilities of propositions reflect the agent's preferences in the light of his total system of beliefs.

Then it is only the complete consistent novels that can be said to have nonprobabilistic values: the desirability of a proposition will be a probability-weighted average of the values of the possible worlds in which it would be true. We have indicated how the agent's preference ranking may (up to a fractional linear transformation) determine a desirability assignment which in turn determines a probability assignment. Moreover, we can go a step further, for the desirability assignment more or less determines an assignment f of nonprobabilistic values to possible worlds. (The qualification "more or less" is required in the light of problem 6 of chapter 6, which illustrates the fact that the values of all the possible worlds that belong to some proposition of probability 0 can be altered arbitrarily without affecting the preference ranking or the assignment of desirabilities to propositions.) The assignments f and *prob* are quite independent, and together determine the assignment *des*.

Just as deductive logic is unconcerned with the question of which possible world is actual, so the logic of decision is concerned neither with the agent's belief function nor with his underlying value function. Rather, it provides a framework within which one can study the relations among various possible belief, value, and desirability functions, and between these and policies for decision making. The framework is normative in much the same way that deductive logic is: it is not put forth as a descriptive psychological theory of belief or value or behavior, but as a useful representation of some very general norms for the formulation and

of Statistical Inference, by L. J. Savage et al. (London: Methuen; New York: Wiley, 1962).

The complete consistent novels of section 12.8 correspond to the maximal sum ideals (ultrafilters) in the Lindenbaum algebra of the agent's language. The interpretation of propositions as sets of novels is a transcription into more colorful language of a device by which one can prove Stone's representation theorem, i.e., "Every Boolean algebra is isomorphic to an algebra of sets": see Paul C. Rosenbloom, *The Elements of Mathematical Logic* (New York: Dover, 1950), pp. 18–27.

In the terminology of mathematical probability theory, the assignment f of nonprobabilistic values to possible worlds (see discussion toward the end of section 2.8) is a Radon-Nikodym derivative of the function *int* (= *prob* × *des*), relative to the measure *prob,* and *des* is the conditional expectation of f: *des A* is the integral of the value function, f, relative to the measure $prob_A$.

For further development of the idea of objectification, see David Lewis, "A Subjectivist's Guide to Objective Chance," in volume 2 of *Studies in Inductive Logic and Probability* (1980); Brian Skyrms, "Resiliency, Propensity, and Causal Necessity," Journal of Philosophy 74 (1977): 704–13; and Bas van Fraassen, "A Temporal Framework for Conditionals and Chance," *Philosophical Review* 89 (1980): 65–108. The last two of these are reprinted in Harper, Stalnaker, and Pearce, *Ifs* (cited in Section 1.8). For Skyrms's work, see his *Causal Necessity* (New Haven: Yale University Press, 1980), chapter IA. There is a survey of such work in section 2 of my "Choice, Chance, and Credence," *Philosophy of Language/Philosophical Logic,* ed. Guttorm Fløistad and G. H. von Wright (The Hague: Martinus Nijhoff, 1981), pp. 367–86 (note that on page 372, line 3 from the bottom, "$c_r = c_p$" should be "$c_r = p_r$").

Appendix
Preference among Preferences

George M. Akrates prefers smoking to abstaining:

$$S \textit{ pref } \overline{S}$$

But he would have the opposite preference if he could, for he prefers preferring abstaining:

$$(\overline{S} \textit{ pref } S) \textit{ pref } (S \textit{ pref } \overline{S})$$

That makes sense. It also makes sense to go further and say that as between his first-order preference for smoking and his second-order preference for abstaining, he prefers the latter:

$$[(\overline{S} \textit{ pref } S) \textit{ pref } (S \textit{ pref } \overline{S})] \textit{ pref } (S \textit{ pref } \overline{S})$$

In these examples, S is the proposition that Akrates is a smoker, and X *pref* Y is the proposition that Akrates prefers truth of X to truth of Y.

I take it that humans (barring infants, psychopaths, . . .) have higher-order preferences and that few if any other animals do, even though other animals may well have first-order preferences. My late cat clearly preferred milk to water on various occasions, but that was surely the end of it. It is not that he was complacent about his preference for milk in a sense in which Akrates, above, is *not* complacent about his preference for smoking (when he prefers smoking, but prefers preferring abstaining). The point is rather that cats are not aware of their preferences in the way in which they are aware of saucers of milk—as objects of desire or aversion, which at least occasionally can be sought or avoided. They are not self-conscious or self-manipulative in that way. But people are, or can be; and therefore in discussing human preferences we want the additional

"Preference among Preferences" first appeared in slightly different form in *Journal of Philosophy* 71 (1974): 377–91.

expressive power that is obtained if we manage to represent preference as an iterable operation that makes new propositions out of pairs of propositions, and not simply as a relation between propositions.

Skepticism about the existence of higher-order preference—or, if you will, skepticism about the need to include higher-order preferences in accounts of human action and motivation—may reflect awareness that for the most part we do not simply choose our preferences, any more than we choose our beliefs, so that direct manifestation of preference in choice is rare when the objects of preference are themselves preferences. But although Akrates cannot simply choose to prefer abstinence on this occasion—and that, after all, is why his preference for preferring abstinence is (uneasily) compatible with his preference for smoking—he can undertake a project of modifying his preferences over time, so that one day he may regularly prefer abstinence, just as now he regularly prefers smoking. The steps toward this desired end may involve hypnosis, reading medical textbooks, discussing matters with like-minded friends, or whatever. But in accounting for Akrates's undertaking of these activities it seems natural to cite his preference for preferring abstinence, just as in accounting for his activities as he flings drawers open and searches through pockets of suits, one may cite his preference for smoking—imagining, in this latter case, that here smoking is not something he can simply choose, because the means are not at hand: he has run out of cigarettes.

There is a traditional view of Akrates's conflict which sees it as a tug o'war between Appetite and Will. Appetite pulls toward smoking. If Akrates were a simple pleasure-seeker, Will would march arm in arm with Appetite, and there would be no conflict. But Akrates would be reasonable, and therefore Will pulls against Appetite. Being weaker than Appetite on this occasion, Will loses the struggle: Akrates (the taut rope) is drawn into smoking, against his will.

So decked out, as a little theory, the traditional way of speaking seems ludicrous. But there are situations in which it is apt to speak of weakness of will, and of strength of appetites for food, sex, cigarettes, or whatever, and where we aptly speak not of preferences but of needs, wants, wishes, and the like. Capitalized and sealed into a tight little language game, Appetite and Will appear to live only by taking in each other's washing; but as part of the full linguistic apparatus that we bring to bear in electing action and in understanding it, these notions have real uses. In particular, we speak of appetites, will, needs, wishes, etc., in making judgments about what people's preferences are.

In *its* tight little language game, the technical term "preference" lives by exchange of washing with the notions of choice, optionality, and judgmental probability. But to make practical use of these concepts, we must bring them into contact with ordinary talk of needs, wishes, preferences, etc., somewhat as we must be prepared to recognize the colors and odors and tastes of things in order to put theoretical chemistry to use. (To get started, we must be able to identify this particular odorless salty white stuff—at least tentatively—as NaCl.) The general idea is that wishes, needs, lusts, etc., are all prima facie signs of *pref* (= preference in the technical or regimented sense) and that, in case of conflict, *pref* is shown by the outcome, which need not be evident in action; e.g., because, of two propositions, that which is preferred true may not be in the agent's power to make true. For the most part, when a proposition of form *A pref B* is thought to be true, the corresponding colloquial statement in terms of preference is taken to be clearly true (on a sufficiently fussy reading), even though it may be thought inappropriate: not what would spring to mind as apt under the circumstances. Perhaps the following adaptation of a remark of Dale Gottlieb's will be helpful here: for the most part, when someone acts in order to make *A* true, and in fact *A pref B* is true, and the only options are making *A* true and making *B* true, the statement that *A pref B* is true will not be satisfactory as a causal explanation-sketch of why the agent acted as he did. Rather, that statement is of some use in suggesting the form appropriate for such a sketch, but the sketch itself will mention needs, wishes, etc., which may be viewed as causes both of the preference and of the action. (Reasons, too, maybe.)

To a first approximation, free action is simply action in accordance with preference. In this sense, Akrates acts freely when he smokes because (= from motives in view of which) he then prefers smoking to abstaining, even though he then prefers abstaining. To be a bit more precise, his smoking is then a free action as long as smoking and abstaining are both options for him, even if preferring to abstain is not. In contrast, consider the case in which Akrates smokes now out of compulsion—against his preference for abstaining (and against his will, in another way of speaking). Here the position is that abstaining is not an option for him, any more than smoking is an option for him when he can get no cigarettes. But perhaps (as Clark Glymour has suggested) action in accordance with preference is not free if the agent would undo that preference if he could. Here as throughout, my aim is not to explicate such common terms as "freedom" and "compulsion" but to show how the notion of higher-order preference can draw no less real distinctions, to much the same effect, in its own terms.

It seems to be only occasionally that preferences are optional, and these occasions may well be seen as fateful. It may well be that Akrates chose his preference for smoking at the age of 14. Perhaps he saw it as part of a larger option: for an adult, masculine, vivid style instead of something prudent or tame or prissy. But ten years and 100,000 cigarettes later, preferring abstention is no longer an option. Early on, each cigarette was a ratification of his choice of preference for smoking, but long since, these acts have become simple *expressions* of well-entrenched preference. Meanwhile, his second-order preferences may have changed, e.g., in response to data about the effect of smoking on health coupled with a sense of having been had, a decade earlier, by the cigarette companies. At this point in his life, Akrates's position would seem to be this:

$$S \; pref \; \overline{S}, \quad (\overline{S} \; pref \; S) \; pref \; (S \; pref \; \overline{S}),$$
$$\mathrm{O}S, \quad \mathrm{O}\overline{S}, \quad \overline{\mathrm{O}(\overline{S} \; pref \; S)}, \quad S$$

where O may be read, "it is optional that." More needs to be said about optionality, soon.

Meanwhile, notice another way in which preferences may be chosen: the way of The Good Soldier who, in obeying an order, acts in accordance with a preference or a set of preferences which stem from the order and from his commitment to adjust his preferences to his orders, but from no wish or desire or need, etc., of the sort that such action would be seen as expressing by someone who took the soldier to be acting on his own, and not in response to an order. The order may be to take a certain hill. There is no question of robot-like response, for the order is not a detailed set of instructions for the placement of his feet, etc.; The Good Soldier will use his wits to make a series of decisions with a view to achieving the required objective, in the light of a background of preferences which may correspond to standing orders about circumstances in which it is acceptable to risk life and equipment in various categories. Adopting a preference on command may well be a matter of preferring in spite of contrary wishes, inclinations, appetites, and fears. Here one may speak of courage, or will power, depending on how smoothly the struggle is won—when it is won. Adopting a preference on command may also be a matter of setting aside standing preferences which, if questioned in civilian life, would be defended in terms of morality or common decency, and which, for that very reason, are not questioned there. Returning to his old life, the soldier may come to see his creditable military performance as acquiescence in monstrous evil:

> And you have to come home knowing you didn't have the guts to say it was wrong. A lot of guys had the guts. They got sectioned out, and on the discharge, it was put that they were unfit for military duty—unfit because they had the courage. Guys like me were fit because we condoned it, we rationalized it. [*Time,* October 23, 1972, p. 34.]

I mention this in order to support the view that following orders is a matter of adopting preferences: the suggestion is that remorse is embittered by a sense of real complicity, for the good soldier acts freely, in accordance with first-order preferences he has freely adopted in accordance with a second-order preference for adopting certain sorts of first-order preferences on command.

If preferences are in the mind, the mind is not always transparent to the mind's eye: one may be unaware of what one's preferences are, in particular cases. One may be in doubt about what they are. And one may be in error in the matter. I take the touchstone to be choice: if *A* and *B* are options, if they are all the options, and if you believe both of those things, then choice reveals preference in that, choosing *A,* you cannot have preferred *B.* But this is far from being a behavioral or operational test. The soldier who, thinking he was about to obey the order to advance, finds himself in flight with empty bowels, may be judged (e.g., by a court martial) to have acted in accordance with a keen preference not to advance; or he may be judged to have acted under compulsion, against his preference to advance. (Will lost the tug o'war with Fear.) On the second reading, advancing was not an option, and the soldier's preference was as he had thought it. On the first reading, advancing was an option, but he was wrong about his preference. Unexpected sexual shrivelings—episodes of impotence or frigidity—are another case in point. Do not Casanova's elaborate preparations to bed the lady make it clear where his preference lies? If so, consummation was, unexpectedly, not an option for him. But on the other hand, does not his most basic indicator of present sexual preference give the lie to his feverish preparations, and show that his real preference is not what he had thought? In such cases, preference and optionality exhibit a sort of complementarity: in order to maintain belief in the one, it is necessary to deny the other.

Nor, in the absence of such ambiguities, need it be clear what preference underlies a publicly observable act. As Donald Davidson has emphasized, actions are events, and events are concrete particulars. But preference relates propositions, not events. If the soldier preferred flight to attack, and fled, this flight is a concrete particular, having no end of properties. (One might describe the course of his flight, e.g., in unlimited detail.) But since he did not attack, there is no concrete particular ("his attack") to serve as the other term of the preference relation, here. When you act in accordance with your preferences, you enact a certain proposition: one of the highest-ranking among those of your options which you believe to be options. You enact the proposition *A* (or, equivalently, you make the corresponding sentence true) by performing some perfectly definite act—an act which can be described as making *A* true, but which can also be described truly as making any number of other propositions true. Among all these, it need not be obvious either to the agent or to an observer, which is the proposition for the sake of the truth of which the act was performed. Here, some may prefer to speak of preferences and choices among *possible* acts, only one of which will prove to be *actual,* viz., the act chosen. But I prefer sentences or propositions as relata of preference. These have the virtue of clarity: ordinary logic provides a satisfactory account of their interrelations, and on top of that account one can construct what I take to be a satisfactory account of preference. But *possible acts* remain to be clarified. I think the burden of clarification is on those who would construct an account of preference in such terms.

If the system of *The Logic of Decision* is to be extended to encompass preference among preferences, it is essential that matters stand as I have argued that they do: that one need not be aware of just what one's preferences are, and that one may have (and, commonly, does have) various degrees of belief between 0 and 1 in propositions of form *A pref B* where "*pref*" refers to one's own preferences. If, on the contrary, degrees of belief in such propositions always or commonly take only the extreme values, 0 or 1, depending on whether the propositions are false or true, there can be no interesting structure of higher-order preferences. The reason is that, in the system of *The Logic of Decision,* the agent must be indifferent between any two propositions in which he has full belief. It would follow that (always or commonly) the agent will be indifferent between pairs of propositions that truly describe his actual preferences. Thus, if he prefers smoking to abstaining, but prefers preferring abstaining, he will be indifferent between those two preferences, simply because he knows that they *are* his preferences.

We wish to leave open such possibilities as

$$[(\overline{S}\ pref\ S)\ pref\ (S\ pref\ \overline{S})]\ pref\ (S\ pref\ \overline{S}) \tag{1}$$

when both relata of the main occurrence of *pref* are true. We have seen that we can do this only if truth, in such cases, does not imply full belief: the proposition (1) will be false—there will be indifference, not preference, between $(\overline{S}\ pref\ S)\ pref\ (S\ pref\ \overline{S})$ and $S\ pref\ \overline{S}$—if *prob,* the agent's judgmental probability function, assigns the value 1 to both propositions. But is this not too high a price to pay? Do we not wish to be able to assert (1) even when the agent is sure that his preferences are as described on the two sides of the main occurrence of *pref*? I think not. Here my reasons are the same as my reasons for holding that, in general, the agent must be counted as indifferent between the truths of *any* two propositions of whose truth he is utterly certain. Preference for truth of one sentence over truth of another can be thought of in a rough and ready way as willingness to pay something in order to have the preferred proposition come true, when the option is between truth of the one and truth of the other. But where the agent is quite sure to begin with that both propositions *are* true, he should not be willing to pay anything to have either of them *come* true. Thus, where the agent is quite sure that (*S*) he will smoke now, and equally sure that ($S\ pref\ \overline{S}$) he prefers smoking now to abstaining now, he must be indifferent between these two. Preference is a practical concept which takes beliefs fully into account in this way. Of course, all of this is compatible with (say) present preference for future preference for abstaining then, over smoking then, where we use two different preference relations, and the sentence about smoking is no longer S. I shall have nothing to say in this paper about the influence of present preference and present action on future preference. (The present discussion of preference is purely synchronic. Thus, a third plausible analysis of the situations of Casanova and The Bad Soldier was omitted above: they might have changed their minds!)

With so much preamble, let us examine various hypotheses about Akrates's situation with a view to seeing how prima facie difficulties may be overcome.

We have noted that, if abstention is not an option for Akrates, there is no difficulty in understanding how he can smoke and yet prefer not to. Here there is no need to invoke higher-order preferences in order fully to describe his situation.

But the case suggested at the beginning is one where Akrates smokes *and prefers to smoke,* but is dissatisfied with his smoking because (a) abstention is an option, and he knows it, and (b) he is dissatisfied with his first-order preferences, and wishes they were otherwise: he prefers preferring abstaining. It is tempting to insist that matters must stand as follows, where, again, OX means that X is an option for Akrates.

$$\begin{array}{ll} \text{(i)} & S\ pref\ \overline{S} \\ \text{(ii)} & S \\ \text{(iii)} & (\overline{S}\ pref\ S)\ pref\ (\overline{\overline{S}\ pref\ S}) \\ \text{(iv)} & \underline{\mathrm{O}\overline{S}} \\ \text{(v)} & \mathrm{O}(\overline{\overline{S}\ pref\ S}) \end{array}$$

The point of asserting (v) is that if preference for abstention were one of Akrates's options, his preference for that option over its denial would seem incompatible with his not taking it, i.e., with his actual preference for smoking over abstaining. But this reason for (v) will not do.

The difficulty is that there may be incompatible options, each of which Akrates prefers to its denial. Then the principle that each option which is preferred to its denial must be true would lead to the conclusion that Akrates makes a logical impossibility true. The rejected principle is this:

(2)
$$\begin{array}{c} \mathrm{O}X \\ \underline{X\ pref\ \overline{X}} \\ X \end{array}$$

Thus, suppose that A and B are sentences which Akrates would like to have agree in truth value—both true, or both false, he cares not which—and that both AB and $\bar{A}\bar{B}$ are options. Setting $X = AB$ in the rejected principle, we have A and B both true, and then setting $X = \bar{A}\bar{B}$ there we also have A and B both false. Concretely, let A = *I shall study Professor Moriarty's treatise on the binomial theorem this afternoon* and let B = *This evening I shall try to explain the binomial theorem to Dr. Watson,* where the laborious business of enacting A would so enhance the probability of success of B as to make AB an attractive option (preferred to its denial), and where it would also be attractive to escape both tasks.

The solution to this difficulty—so I think—is to compare X with all other options, and not simply with $\bar{X}$ (when that happens to be an option, which it need not be). Thus, an improvement over the rejected principle would be this, in idiomatic English:

> A proposition must be true if making it true is an option preferred to every other.

It is simply not good English to speak of propositions as options, although I shall continue to do so for fluency when the propositions are represented in logical notation. OX is best read, "Making X truc is an option," and that may be taken as the intent of my barbarous "X is an option."

To a first approximation, an option is a proposition which the agent can be sure is true, if he wishes it to be true. (Thus, $A \vee \bar{A}$ is always an option, trivially.) But there may be options in this sense which the agent does not know are options, as when he fails to realize that a door is unlocked, or that his car will start, or that his love is reciprocated, or that if he holds out one more day, the pain of withdrawal will begin to ease. Perhaps such unrecognized options should not be regarded as options at all, or perhaps some should but others should not. At any rate, the principle seems to need further restatement, perhaps as follows:

> A proposition is true if making it true is a recognized option preferred to every other recognized option.

I shall continue to use the term "option" in the broad sense in which not every option need be recognized as such. Problem: how shall we characterize the recognized options? Can this be done in the terms we have already deployed, or is a new primitive term needed? One might think of defining recognition as full belief in a truth: where *prob* is the agent's judgmental probability function, one might take joint truth of OX and *prob* $(OX) = 1$ to be necessary and sufficient for X to be a recognized option. The question whether *prob* $(OX) = 1$ can be settled by examining the agent's preference ranking, in which such sentences as OX and $O\bar{X}$

are presumed to occur. (For details, see section 7.4.) On this reading, recognition is definable in terms already at hand, but it is unclear that this is an adequate reading. Must one be as sure that OX as one is that $2 + 2 = 4$ in order to be said to recognize that making X true is an option? The basic difficulty here is a familiar one: lack of fit between ordinary talk of knowledge, belief, and recognition on the one hand, and such notions as judgmental probability and preference on the other. The principle we have been elaborating becomes a rule of very narrow scope when recognition is defined in terms of *prob* as above. On the other hand, if we use the colloquial form of the rule, giving "recognized" its colloquial reading, we have a looser principle of broad scope, extraneous to the theory of preference, but serving to link that theory with ordinary ways of speaking. Perhaps the principle is best taken in its colloquial form. One might still include otionality as a concept within the theory of preference, and might use the colloquial principle as a guide in dealing with particular applications of the theory.

We have roughly characterized an option as a sentence which the agent can be sure is true, if he wishes it true. Presumably the reference to wishing, here, can be replaced by a reference to preference. Can we then define optionality in terms of preference? Any such definition would seem to require consideration of whether the optional sentence would be true if the agent's preferences and beliefs were other than they are. The Small Girl in Lewis Carroll's joke (see W. W. Bartley III in *Scientific American* 227 [1972]: 39) has taken a faltering step in that direction:

> "I'm so glad I don't like asparagus," said the Small Girl to a Sympathetic Friend, "Because, if I did, I should have to eat it— and I can't bear it!"

Such a definition has been suggested by David Lewis, using the notion of a *partition,* viz., a set of propositions, no two of which can be true and at least one of which must be true. Such a set is a *partition of options* if and only if the following holds for each member: if it were preferred to every other member then it would be true. Now a *basic option* is defined as any member of any partition of options, and an *option* is defined as anything implied by a basic option.

This definition of O will serve if each option really is implied by some basic option; and that seems to be the case. Thus, one of my options is dying before the age of fifty years, e.g., by my own hand, but that is no basic option, for, I regret to say, its denial ("Live at least fifty years") is nothing I can ensure. But it is an option according to Lewis's definition: there is a partition of options ("Jump out the window"; "Don't"), one member of which implies it. Lewis's analysis seems right. I am still a bit queasy about counterfactuals, but Lewis is not, and perhaps he is right, not to be.

In conclusion, let us examine a set of preferences more contorted than any attributed to Akrates heretofore. Suppose that Akrates's preference for smoking, like the smoking itself, is the outcome of a preferential choice, and that nevertheless, he prefers not preferring to smoke. Thus, each of the following is true:

$$S, \qquad S \textit{ pref } \overline{S}, \qquad (\overline{S \textit{ pref } \overline{S}}) \textit{ pref } (S \textit{ pref } \overline{S})$$

Furthermore, the first two of these propositions are optional, as are their denials. Finally, Akrates believes (even, with judgmental probability 1) that these four are options and are all the options. Can we consistently suppose all this? I think so.

An air of paradox lingers round the claim that $S \textit{ pref } \overline{S}$ is true and optional, whereas its denial is preferred to it; but those are not all the options. Akrates's situation is intelligible enough if we inquire into the relative positions of S and $\overline{S \textit{ pref } \overline{S}}$ in his preference ranking, and find that the former is higher, e.g., because we have one of the following two configurations, in which better-liked options are higher:

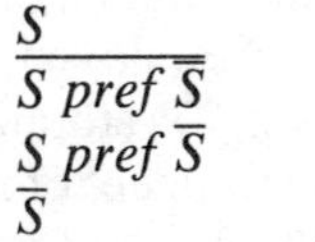

(1)
$$\begin{array}{l} S \\ \overline{S \textit{ pref } \overline{S}} \\ S \textit{ pref } \overline{S} \\ \overline{S} \end{array}$$

(2)
$$\begin{array}{l} \overline{S} \\ \overline{S \textit{ pref } \overline{S}} \\ \overline{S} \\ S \textit{ pref } \overline{S} \end{array}$$

In each case, the fact that S is above $\overline{S}$ shows that $S \textit{ pref } \overline{S}$ is true, even though the denial of that option is preferred to it. In each configuration, the top entry is true perforce, being the best of the (recognized) options. But in each case, one of the other entries is also true, and in neither case is that one the best of the remaining options. (In the second case, it is the worst that is chosen, along with the best!) It might seem plausible to explain the consistency of the first configuration by pointing out that there, the intensity of Akrates's preference for smoking over abstaining is greater than that of his preference for not preferring smoking, over preferring it. But that need not be the case in the second configuration. The common, telling feature is simply that his desire to smoke is at least as intense as his desire not to prefer smoking to abstaining: in fact, $S \textit{ pref } \overline{S \textit{ pref } \overline{S}}$. Were *that* preference reversed, inconsistency would ensue, e.g., in the following ranking:

(3)
$$\begin{array}{l} \overline{S \textit{ pref } \overline{S}} \\ S \\ \overline{S} \\ S \textit{ pref } \overline{S} \end{array}$$

If these four are taken to be options and to be all the options, the topmost must be chosen; but then the order of the middle two cannot be as shown.

Perhaps enough has now been said, to indicate the nature and objectives of the theory of higher-order preference. Like the theory of first-order preference, it simply refuses to countenance as preferences, rankings that are intransitive or have certain other failings—human though those failings be. But the important point is that the higher-order theory does countenance various other failings—or misfortunes, or conflicts, or tensions, or "contradictions" in some Hegelian sense. It gives us a canvas on which to paint some very complex attitudinal scenes, from life. That one cannot paint intransitive preference rankings on that canvas makes it all the more interesting that one can paint poor Akrates there, in the various postures we have seen above.

Notes and References

"Preference among preferences" first appeared in the *Journal of Philosophy* pretty much in its present form. Apart from acknowledgments below, thanks go to Fabrizio Mondadori, Jay Rosenberg, James Cornman, Edwin McCann, and, especially, Brian Chellas, for useful suggestions.

The notion of higher-order desires had been deployed by Harry Frankfurt, in a paper I read tardily after finishing "Preference among preferences": see his "Freedom of the Will and the Concept of a Person," *Journal of Philosophy* 68 (1971): 5–20, where the ability to form second-order desires is taken to be the mark of personhood.

Donald Davidson's papers on actions, events, reasons, and causes have been handily collected in *Essays on Actions and Events,* (Oxford: Oxford University Press, 1980).

For one account of possible acts, see Savage, *Foundations of Statistics* (1954): he takes the relata of preference to be certain entities which he calls "acts" and which might better be called "possible acts." These entities are functions from the set of all possible worlds to a set of entities called "consequences." The consequences may, but need not, be hedonic states of the agent. Here, one surely never knows just what act one is performing; for to know that, one would have to know how the act would turn out in every possible world. If possible acts are such as these, they cannot be the objects of choice. These points are elaborated in my "Frameworks for Preference," in Balch, McFadden, and Wu, *Essays on Economic Behavior under Uncertainty* (1974).

The principle (2) runs afoul of the difficulty that there may be incompatible options, each of which Akrates prefers to its denial. One might try to avoid the difficulty by drastically weakening the principle so as to require only that an option be false if its denial is preferred to it:

(2′)
$$\frac{\begin{array}{c} OX \\ \overline{X}\ pref\ X \end{array}}{\overline{\overline{X}}}$$

The underlying thought ("the Anscombe-Kanger principle") is that *good is good enough.* But in the text I take another way out, in which connections among preference, optionality, and action remain relatively tight.

The following curious feature of David Lewis's definition of optionality has been pointed out by Joseph Ullian. Consider the ranking (2) above, i.e.,

$$\begin{array}{c} S \\ \overline{P} \\ \overline{S} \\ P \end{array}$$

where P is the proposition $S\ pref\ \overline{S}$. Since $\{P, \overline{P}\}$ is a partition and both P and $\overline{P}$ are options, one might expect $\{P, \overline{P}\}$ to be a partition-of-options in Lewis's technical sense; but it cannot be, for $\overline{P}$ is false although it is preferred to P. To this Lewis replies that P and $\overline{P}$ cannot be *basic* options in this case. Instead, the basic options must be SP, $S\overline{P}$, $\overline{S}\overline{P}$, and $\overline{S}P$, in descending order of preference. But, as Lewis conjectures, the situation remains disturbing in that although the basic option SP is actualized, Akrates cannot clearly realize that, for one can show that the given preference ranking is possible only when Akrates's degree of belief in SP is less than 1/2. Reasons: for P to be at the bottom of the ranking, the probability-weighted average of the best and worst basic options must be less than that of the two lowest. From this, with some labor, one can get $prob\ SP < prob\ \overline{S}\overline{P}$, whence the result follows.

As noted above, the present theory "simply refuses to countenance as preferences, rankings that are intransitive or have certain other failings—human though those failings be." My thought is that someone who says he prefers A to B and B to C but not A to C is simply mistaken about his preferences. Others prefer to say that the theory imposes an idealization, and is false of much actual preference. Still others (e.g., Amélie Rorty) hold that intransitivities may be quite in order, e.g., when A is preferred to B in one respect (or, under one description) and B is preferred to C in (or, under) another. But throughout, I am concerned with preference *all things considered,* so that one can prefer buying a Datsun to buying a Porsche even though one prefers the Porsche qua fast (e.g., since one prefers the Datsun qua cheap, and takes that desideratum to outweigh speed under the circumstances). *Pref* = preference *tout court* = preference on the balance.

The problem of providing a semantics for higher-order preference seems fairly straightforward, when optionality is bracketed. The basic point (suggested by David Lewis ca. 1966) is that nothing in the system of *The Logic of Decision* requires us to suppose that the terms of the preference relation are naturalistic propositions. They may equally well be propositions about the agent's preferences. In the case of propositional logic with *pref* and *ind* (for "indifference") as the only non-truth-functional connectives, the set L of sentences under consideration may be taken to be the closure of a countable set of atomic sentences under connectives for denial, conjunction, preference, and indifference (i.e., overbar, juxtaposition, "*pref*," and "*ind*," which are now taken to apply to sentences, not propositions). A *model* of L is determined by the following five items:

(1) A nonempty set W of "possible worlds"
(2) A Boolean algebra of subsets of W, viz., "propositions"
(3) For each w in W, a probability measure P_w on the algebra
(4) For each w in W, a (utility) function u_w which assigns a real number $u_w(w')$ as value to each argument w' in W in such a way that the conditional expectation $E_w(u_w|A)$ of u_w on A relative to the probability measure P_w exists for each A to which that measure assigns positive probability. Note that in computing the conditional expectation, w is held fixed: $E_w(u_w|A)$ is the probability-weighted average of the values $u_w(w')$ that the function u_w assigns to arguments w' in A, where the weights are determined by the fixed probability measure P_w. The resulting conditional expectation is a function of w but not of w'. Thus,

$$E_w(u_w|A) = (1/P_w(A))\int_A u_w dP_w \,.$$

(5) For each S in L, an element $M(S)$ of the Boolean algebra ("the proposition expressed by S") satisfying the conditions:
   (a) $M(\overline{S}) = W - M(S)$
   (b) $M(ST) = M(S) \cap M(T)$
   (c) $M(S \text{ } pref \text{ } T)$ = the set of all those worlds w which satisfy the inequality $E_w(u_w|M(S)) > E_w(u_w|M(T))$, where the inequality is taken to fail if either side is undefined
   (d) $M(S \text{ } ind \text{ } T)$ = the set of all those worlds w which satisfy the condition $E_w(u_w|M(S)) = E_w(u_w|M(T))$, which is taken to fail if either side is undefined

I am indebted to Zoltan Domotor for help in getting this right, if it is right.

The heart of the construction is in (5) (c) and (d), which ensure that "*S pref T*" (or "*S ind T*") will be true in world w if and only if S is higher than T (or S is at the same level as T) in the preference ranking determined by the agent's values (u_w) and beliefs (P_w) as they would be in that world.

In general, S is true in world w if and only if w belongs to $M(S)$. Truth unqualified is truth in the real world, viz., an unknown element r of W. One might think of adding a sixth item to the five that are taken to determine a model: (6) A distinguished member r of W which satisfies the condition that for each S in L to which P_r assigns positive probability, $E_r(u_r|M(S))$ is the P_r-weighted average of the values assumed by $E_w(u_w|M(S))$ as w ranges over W. But perhaps this condition, if apt, should be imposed on every element r of W. Validity in a model is truth in every member of the set W for that model. Universal validity is validity in every model. In terms of this semantics one might discuss probabilities of probabilities, probabilities of preference, and so forth.

Index